Forensic Biomechanics

א

Using Medical Records to Study Injury Mechanisms

Other SAE titles of interest:

Neck Injury:
The Use of X-Rays, CTs, and MRIs to Study Crash-Related Injury Mechanisms
By Jeffrey A. Pike
(Product Code: R-268)

Vehicle Accident Analysis and Reconstruction Methods
By Raymond M. Brach and R. Matthew Brach
(Product Code: R-311)

Automotive Safety Handbook, Second Edition
By Ulrich Seiffert and Lothar Wech
(Product Code: R-377)

Role of the Seat in Rear Crash Safety
By David C. Viano
(Product Code: R-317)

Forensic Biomechanics

א

Using Medical Records to Study Injury Mechanisms

Jeffrey A. Pike

SAE International

Warrendale, Pa.

For permission and licensing requests, contact:

SAE Permissions
400 Commonwealth Drive
Warrendale, PA 15096-0001 USA
E-mail: permissions@sae.org
Tel: 724-772-4028
Fax: 724-772-4891

Library of Congress Cataloging-in-Publication Data

Pike, Jeffrey A.
Forensic biomechanics : using medical records to study injury mechanisms / Jeffrey A. Pike.
p. ; cm.
Includes bibliographical references and index.
ISBN 978-0-7680-1801-1
1. Crash injuries. 2. Traffic accidents. 3. Medical records. 4. Human mechanics.
I. Society of Automotive Engineers. II. Title.
[DNLM: 1. Wounds and Injuries--physiopathology. 2. Accidents, Traffic.
3. Biomechanics. 4. Forensic Medicine--methods. 5. Medical Records.
WO 700 P635f 2008]
RD96.6.P552 2008
617.1'028--dc22 2008001949

SAE International
400 Commonwealth Drive
Warrendale, PA 15096-0001 USA
E-mail: CustomerService@sae.org
Tel: 877-606-7323 (inside USA and Canada)
724-776-4970 (outside USA)
Fax: 724-776-1615

ISBN 978-0-7680-1801-1

SAE Order No. R-379

Printed in the United States of America.

To Debbie
and the "kids,"
Stacy, Adam, Jessica, Blair, and Emily

Contributors:

John Cavanaugh, M.D.
Professor, Biomedical Engineering
Wayne State University
Detroit, Michigan

David B. Weiss, M.D.
Orthopaedic Traumatologist
Orthopedic Surgery Associates
Ann Arbor, Michigan

Contents

Foreword

The practice of forensics is traceable back to ancient Greek society, which regarded forensics—the use of reasoned discourse in public life—as central to democracy (American Forensics Association, 2008). The roots of biomechanics also can be traced back to this time frame, wherein anatomists studied the structure of the body and later sought to quantify its characteristics.

In our time, biomechanical studies are an integral part of designing and developing various safety systems (e.g., seat belts, airbags), and indeed, federal mandates require regulatory agencies to determine and compare the anticipated societal costs and safety benefits of proposed safety rules (Pike, 1990). Forensic biomechanics can provide clinicians, government regulators, product developers, safety researchers, and triers of fact with an additional methodology for applying reasoned discourse to their decision-making process.

My vision of success is that the methodology presented here will help those in the field to determine how people are injured and how they are not injured, and thereby will assist in developing safer products and environments and will help to develop more effective public policy regarding various safety issues. Furthermore, the methodology will provide clinicians with a methodology to help them determine whether they can rely on the histories that have been provided to them in helping to guide medical care, when supplemental information is needed, and when the histories should be viewed with skepticism.

I hope my vision will prove to be myopic, in that readers will find new ways (that I did not foresee) to utilize and build upon this material and to further contribute to mitigating injury and reducing fatality.

I look forward to receiving reader feedback and can be contacted at jp@BCIconsult.com.

Acknowledgments

In the Acknowledgments section of another book, the author expressed the opinion that the experience of writing two books had been so rewarding that he began thinking in terms of book "number three." Well, that author was me, and about six years and several drafts after book number two, here is book number three.

Much of the material in this book has been adapted from lectures that I have given through the SAE International seminar program, at universities, and at other technical forums. The attendees at these events provided much useful feedback and helped me to identify the information that would be most useful to include in this volume.

I am especially appreciative of the efforts made by the reviewers. Although very busy, they were quite generous with their time and expertise. They taught me a great deal, not only about the subject matter, but also about graciously helping others.

Patient privacy, of course, is a priority, and in the quest for such, identifying information has been changed or deleted, and all dates and years have been changed. Histories also have been changed, and radiographic images have been selected and/or modified to maximize their pedagogic value. (In many real-world instances, injuries are not as easy to detect.)

I hope that readers of this book will contribute to a decrease in injuries and to improved care for those who are injured.

Finally, in regard to book number four (about which there have already been several inquiries), I'd like to quote the sage who said, "It's difficult to tell the future—especially in advance."

Introduction

This book is intended to be read by a wide range of professionals, with the ultimate goal of having them work together more effectively. The underlying belief is that the more we know and can share about how injuries are caused (and how they are not caused), the better we can allocate society's resources to avoid or mitigate injuries and to respond in an improved manner when injuries do occur.

The book seeks to achieve this by improving the utility of injury biomechanics accident reconstruction and field studies by expanding their scope and making the insights gained from such studies more widely applied. This book discusses medical records, how to read them, the types of information they contain, and the types of insight they may provide regarding restraint usage and deployment, injury mechanism, injury severity and outcome, and the effects of prior medical conditions. Annotated case studies provide examples of real-world use of terminology and abbreviations, with comments to integrate and provide a fuller understanding of the context in which they may be applied. Advanced case studies provide the opportunity for additional self-study and/or class exercises. A glossary provides an annotated list of some of the more common abbreviations that may be encountered. Several tables of figures provide composite lists of the medical records arranged by body region, imaging type, and injury type. Finally, several sample expert reports are provided to give additional insight into the wide range of forensic biomechanical issues that may be addressed by using medical records.

Background

Discussion

The study of injury causation and injury mitigation is becoming an ever-expanding interdisciplinary field. The purpose of this book is to help practitioners of the various disciplines speak a common language and to make safety professionals more aware of the types of information that may be available in medical records and the types of issues that this information may be used to address, both directly and indirectly. The book also will provide clinicians with a methodology for obtaining improved patient histories, to aid in their treatment decisions. The book includes case studies, with annotated notes and additional discussion. The case studies provide examples of how medical records may be used to address impact severity and direction; the effect of intoxication and other pre-existing conditions on crash avoidance, injury severity, and injury outcome; restraint usage; airbag deployment and non-deployment; and injury mechanisms, both those that involve occupant impacts with the vehicle interior (the so-called second collision) and those that do not. Note that vehicle occupants also may be injured outside the vehicle, either as a result of ejection or a fall during vehicle egress, and the cause of the fall could be internal (e.g., dizziness) or external (e.g., slippery surface).

Some of the topics that may be addressed using medical records (e.g., restraint usage and impact severity) also may be addressed by vehicle inspection and/or the newer technology of "black boxes," that is, electronic data recorder (EDR) data retrieval and interpretation. Note that although EDRs represent a newer technology, they are not necessarily the gold standard with respect to addressing restraint usage and impact severity. For example, the EDR may contain data from a previous event rather than the event being studied or, if the vehicle was airborne, may have recorded inapplicable speed information. Biomechanical analysis also can provide clinicians with an additional tool for critical analysis of the history provided to them, to better determine which parts of the history should be used as is, which parts require more detail, and which parts should be discounted, as part of diagnosis and treatment decisions. In general, the more different methodologies that can be utilized and results compared for consistency, the higher the confidence in the conclusions and the more likelihood of identifying inconsistencies to be addressed.

The book also will discuss pre-existing conditions and their role in motor vehicle accidents (MVAs), both with regard to injury causation and injury outcome. Note that in the general population, instances of a patient having had a previous injury that was either similar or related are rare. However, a previous injury often places an individual at increased risk for future injury; therefore, in the injured population, "repeat injuries" are not nearly as rare. Furthermore, the previous injury may have occurred at a site of already reduced strength (e.g., osteoporosis). Thus, that same location would be at risk for future injury as well. If a prior injury were the result of risk-taking behavior that persists, this also would tend to place the individual at increased risk of future injury. Finally, medications taken to treat a previous injury or condition may affect driver alertness or occupant frailty. The book also includes a brief glossary of some of the more common medical record abbreviations.

The records may be organized into six groups: (1) narratives, (2) imaging, (3) tests, (4) field reports, (5) pre-existing conditions, and (6) case studies. Narratives include emergency medical services (EMS) reports, emergency room (ER) reports, surgical reports, other hospital records, on-going progress notes, specialist reports, and autopsy reports. Imaging includes x-rays, computed tomography (CT), and magnetic resonance imaging (MRI). Tests include blood, neurophysiological, and neuropsychological tests. Field reports include traffic accident reports (TARs) and vehicle information, including restraint usage. Pre-existing conditions include disease, injury, intoxication, and drug abuse. When pedagogically desirable, more than one figure will be devoted to a single record. Typically, the first figure will provide an overview of the document, and then subsequent figures will "zoom in" on the section being discussed.

Anatomy Concepts

The following discussion is intended to provide a brief overview of some aspects of anatomy that will be useful in the remainder of the text. It is intended to be neither exhaustive nor exhausting, but rather to provide some useful background to those relatively new to the field. Much of the following is derived from Pike, 1990. Those desiring a more detailed discussion are referred to that text.

The Thorax

The sternum or breastbone is in the center front of the chest and is composed of three bones: the manubrium on top, the gladiolus or sternal body in the center, and the xiphoid process at the bottom. A series of curved bones, the ribs, connects the sternum in front and the spinal column in back (Figures A and B). Usually, there are twelve ribs on each side. The first seven ribs, that is, the top seven ribs starting at the upper end of the manubrium and going downward, connect, via cartilaginous extensions, directly to the sternum. Ribs 8, 9, and 10 connect to the ribs immediately above them, and ribs 11 and 12 (the bottom two ribs), or "floating ribs" as they sometimes are called, do not attach to other ribs, but rather are attached to muscles in the abdominal wall.

Although the ribs frequently are thought of as part of the chest, and indeed the ribs are a very important part of the chest structure, the lower ribs do extend into the abdominal area. In fact, these lower ribs cover part of the abdominal organs and thereby provide the abdominal organs with some protection. Located within the chest region of this bony framework are

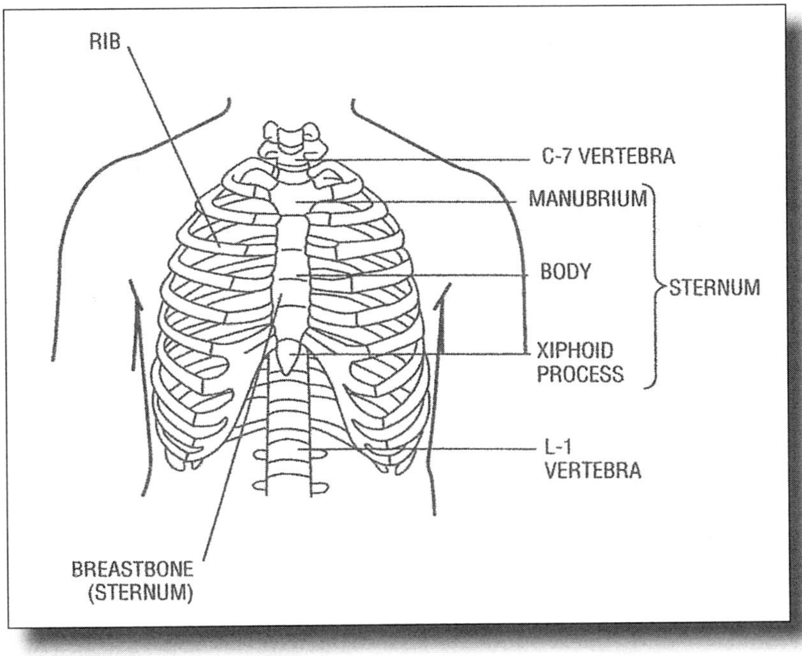

Figure A *The bony thorax—anterior view. (From Pike, 1990, Figure 2-19. Published by SAE International. Used with permission.)*

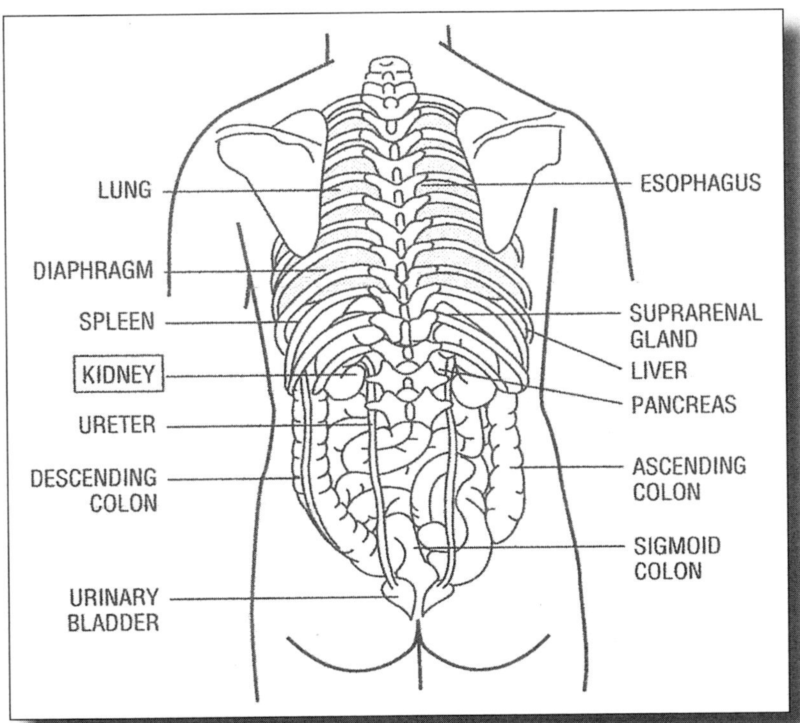

Figure B *The torso—posterior view. (From Pike, 1990, Figure 2-20. Published by SAE International. Used with permission.)*

the left and right pleural cavities, each containing one lung, and the region between them, the mediastinum, which includes the heart. The parietal pleura lines the inside of the ribs, and the visceral pleura lines the lung. Although these two pleurae are quite close to each other, there is a potential space between them, and this space is referred to as the plural space. (As will be discussed later, these pleurae and the space between them are central to certain important injuries.) The mediastinum contains a group of structures including the heart, the major blood vessels going to and from the heart, and the trachea (windpipe) and esophagus (gullet).

As already discussed, the bony frame provides much of the protection for the thorax. In spite of the protection offered by the bony framework, the thoracic contents and, indeed, the bony framework itself may be injured if subjected to trauma that exceeds their tolerance limits. A blunt chest impact is typified by a relatively flat, blunt object (e.g., the steering wheel) striking (or being struck by) the chest without penetrating the chest surface. Depending on variables such as the amount of ribcage compression, the location of the compression on the ribs, the velocity of the compression, and the age of the victim, the ribs may merely deform and then return to their original position, or the ribs may fracture (break). If a fracture occurs, then there is the additional risk that one or both rib edges at the fracture site, which generally are somewhat jagged, may move either outward, until the skin is ruptured, or inward, possibly injuring the various soft tissues near the rib. In those rare instances when the rib fragment moves outward until the skin is broken, the fracture site is described as communicating with the outside environment, and the fracture is classified as an open or compound fracture.

If blood vessels in the thorax are injured, in addition to the usual problems associated with bleeding, there is the potential hazard of blood accumulating within the chest cavity between the two layers of pleura (Figure C). When blood starts to accumulate in the chest cavity, it is referred to as hemothorax. When air accumulates inside the chest cavity (not, of course, inside the lung, as it does during normal breathing, but rather in the pleural space, outside the lung), it is referred to as pneumothorax. The accumulation of either air or blood outside the lung tends to compress or collapse the lung and thereby interfere with the lung's ability to fill with

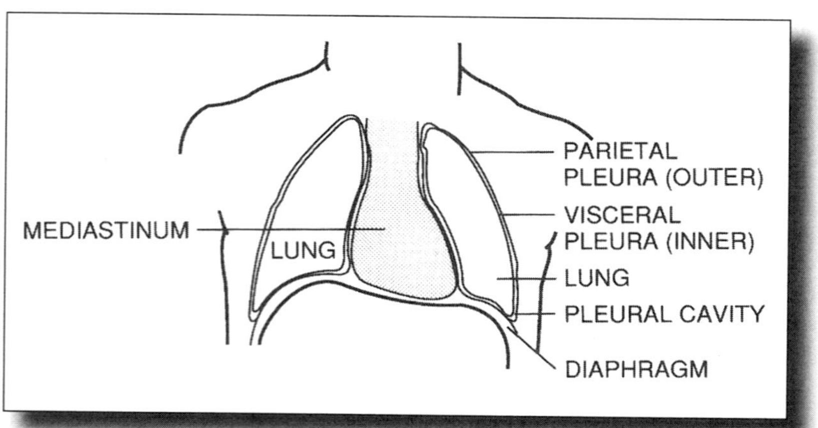

Figure C *The thoracic cavity. (From Pike, 1990, Figure 2-23. Published by SAE International. Used with permission.)*

air and hence with the patient's ability to breathe. Pneumothorax and hemothorax can occur on one side of the chest, that is, affecting one lung, in which case it is referred to as unilateral or single hemothorax or pneumothorax. Or it can occur on both the left and right sides of the chest, in which case it is referred to as bilateral or double hemothorax or pneumothorax (or, if both, hemopneumothorax).

Chest compression also may injure the lung, the heart, the major veins going to the heart, and the major arteries going from the heart. (These veins and arteries are referred to as the great vessels.) These injuries may include a contusion, bruising, or, in the extreme, a rupture. It is important to note that depending on the particulars of the specific accident, the ribs may compress sufficiently to injure some of the soft content of the chest without being fractured themselves. For example, the pleura or the lung might be bruised by a brief slap-like contact from a compressed rib, although the rib undergoes a temporary plastic deformation. That is, the rib does not break or exhibit permanent deformation, but rather quickly returns to its original position. A bruised lung, heart, or great vessel is not an insignificant injury.

Even if surrounding soft tissue is not damaged, rib fracture, especially if several ribs are fractured, can be a serious injury in its own right. If enough adjacent ribs are fractured, a condition referred to as flail chest may occur, and this can severely compromise respiration. Normally, when the diaphragm moves downward, it creates a partial vacuum within the chest. This pressure differential between the outside atmosphere and the air inside the chest causes air to enter via the nose and into the lungs. If injury produces flail chest, the chest wall loses its structural integrity and ceases to function in an integrated fashion during inhalation and exhalation. This also may be referred to as a reciprocating chest injury or as paradoxical breathing. This is because when the patient tries to inhale, although he is trying to raise and expand his chest as in normal inhalation, a part of his chest, the so-called flail segment, will fall inward, thereby moving in the opposite direction. In addition to being extremely painful, the flail chest may make it difficult or impossible for the normal exchange of gases to occur in the lung (Mullins, 1987).

The Spine

The bony framework of the spinal column is composed of bones of various shapes and sizes. These include a series of ring-like vertebrae stacked on top of each other to form a column-like structure, the vertebral column. The channel bordered by the opening in the center of the vertebrae contains the downward extension of the brain stem, which leads to the spinal cord. Spinal cord injuries are of particular concern because they may result in loss of movement and sensation. Furthermore, because the extent of such losses depends on the location of the injury, it is important to be able to accurately specify regions of the spinal cord. A shorthand system frequently used to specify location of spinal cord injuries designates each of the vertebrae using the first letter of the name of the region in which the vertebra is located. Moving inferiorly, that is, downward from the base of the skull toward the tailbone, the areas are named cervical, thoracic, lumbar, sacral, and coccygeal (Figure D). Each individual vertebra then is designated by referring to the first letter of the region (e.g., "T" for thoracic), and, starting with the uppermost (most superior) vertebra in that region, numbering the vertebrae

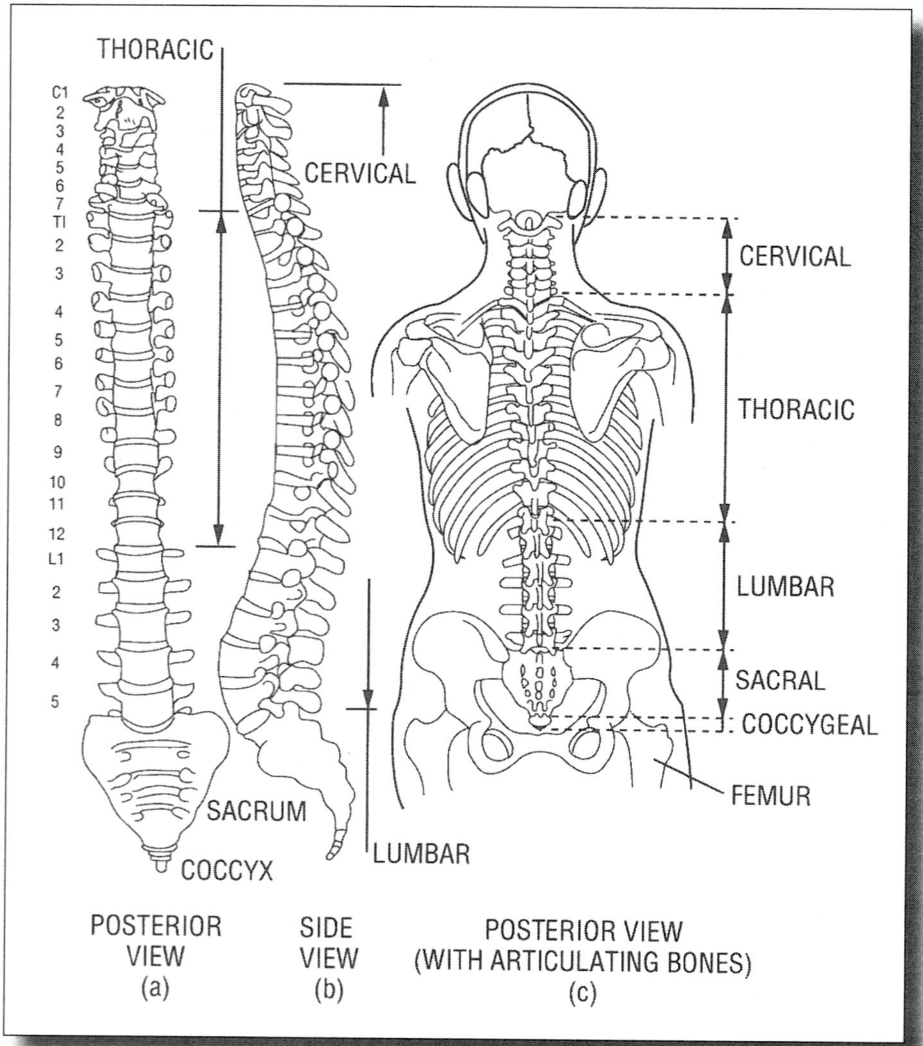

Figure D *The spine. (From Pike, 1990, Figure 1-8. Published by SAE International. Used with permission.)*

consecutively until the lowest (most inferior) vertebra in that region has been named. Thus, the vertebrae of the cervical region are designated as Cl, C2, and so on, down through C7. (Cl is at the head-neck junction, and C7 is at the neck-shoulder junction.) Examples of the use of this nomenclature would be to describe a vertebral fracture "of C5" or a spinal cord injury "at C5." The vertebra immediately below C7 is the first vertebra of the thoracic region; therefore, it is designated as T1. The vertebra immediately beneath this is T2 and so on, until T12. Immediately beneath T12, there is the first vertebra of the lumbar region, which is designated as L1. Beneath this are the remaining lumbar vertebrae, which are designated as L2 through L5. Below L5 is the sacrum, and at the very tip of the sacrum, the coccyx.

At birth, the sacrum is composed of individual bones; however, by adulthood, they have fused into a single bone. Although the constituent vertebrae have fused, they are still discernable. Therefore, a similar shorthand nomenclature may be used with the sacrum—either to designate actual distinct segments in the immature sacrum or to designate regions of the fused sacrum. In either ease, the designations are S1 through S5.

The next region, the coccyx, is composed of vertebrae that are rudimentary, in that they do not have all the sharply defined surfaces of the vertebrae of other regions. There usually are four coccygeal vertebrae. Similar to the sacral vertebrae, the coccygeal vertebrae, even when fused, remain discernable and may be designated as Co1 through Co4. Although four coccygeal vertebrae are most typical, three or five coccygeal vertebrae also may be encountered. Regardless of which of these formations is present in a given individual, the same basic top-to-bottom system would be used to assign labels to them. (Note: The abbreviation CCX sometimes is used rather than Co to designate a coccygeal vertebra.)

When viewed from the side, the vertebrae of a normal spine for an individual in the anatomic position (basically standing, facing forward, arms at sides) present three major curves corresponding to the cervical, thoracic, and lumbar regions. The cervical and lumbar regions display a forward-pointing curve (lordosis), and the thoracic region displays a rearward-pointing curve (kyphosis) (Figure D). Spinal injuries may run the gamut from very minor to very severe. In some cases, there may even be doubt that an injury has occurred.

In some instances, a patient may complain of a so-called "whiplash" injury, with complaints of pain, typically in the neck region, but no structural damage depicted on x-rays and/or other imaging. In these situations, the complaints often resolve within a few days or weeks. When the complaints last longer than this and are in the absence of objective correlates of injury, the caregiver may conclude that the complaints are not related to a recent minor trauma. In other instances, radiographic findings (e.g., a protruding or ruptured disc) may be assigned incorrectly to a minor trauma, when, in fact, these may have been incidental findings of pre-existing conditions (e.g., disc protrusions, disc ruptures) (Gorman, 1997; McRae 1956). Indeed, cervical disc protrusion is a common incidental finding in asymptomatic people who are older than 30 years of age. The incidence increases with age and often includes spinal cord impingement and spinal stenosis (Banks, 2000). Similarly, side-view (lateral) radiographic images that display curvature that deviates from the norm may merely reflect variations among the population or an unusual position at the time of imaging. For example, "cervical straightening" often is an incidental finding and not indicative of trauma (Harris, 1987; Helliwell, 1994; McRae 1956).

Also, an anterior displacement of one vertebra with respect to the vertebra beneath it (spondylolisthesis) is present in approximately 5% of the population without any symptoms. If found on an x-ray taken after a minor trauma, the condition may not relate to the trauma at all. In addition to the pre-existing spondylolisthesis, lumbar imaging, taken after trauma, may show a spondylolysis (separation) at the isthmus (the pars interarticularis) of L5. Typically, this separation is a long-standing condition, that is, a stress fracture or structural defect associated with repeated microtrauma by participants in sports such as tennis. In the absence of an intact isthmus, the connecting tissue assumes the additional role of maintaining vertebral position, and over time, this connecting tissue becomes weakened and permits vertebral displacement. A recent study of asymptomatic athletes found that approximately 85% had various abnormalities on lumbar MRIs, including bulging discs, degenerative bone growths (osteophytes), spondylolysis, and spondylolisthesis (Alyas, 2007).

Most spinal column injuries can be categorized as a sprain, disc disruption, or vertebral fracture (Huelke, 1979). (These injuries may or may not involve the spinal cord and spinal nerve roots.) As will be discussed next, these three categories are by no means mutually exclusive (Bohlman, 1986). A sprain basically refers to the stretching of soft tissue beyond its elastic limit (that is, it will not return to its original shape and size, although the stretching is discontinued). The soft tissue involved most often is a ligament or facet joint capsule (that is, the fibrous structure enclosing a facet joint). Soft tissue injuries may run the gamut from minimal stretch with no bony displacement to severe tears with significant concomitant bone displacement.

It is the spinal cord, contained within the framework of the bony spine, that is of ultimate concern with regard to injury. When the spinal cord is damaged, paralysis and loss of sensation may occur. Thus, vertebral fracture (i.e., a cracking or breaking of one or more of these bones) per se is not sufficient to cause neurological deficit. (Indeed, approximately 60% of cervical fractures have no associated neurological injury [Mahoney, 1988].) Rather, it is the potential of a fractured or dislocated vertebra to cause injury to the spinal cord that is of major concern.

Similarly, a major concern regarding a torn or ruptured disc is whether or not the jelly-like middle of the disc (the nucleus pulposus) is extruded rearward to the extent that it presses on a spinal nerve (a branch of the spinal cord) or possibly even presses on the spinal cord itself (Figure E). When part of the disc material presses against surrounding soft tissues and possibly injures the spinal cord or spinal nerves, the extreme pain and disability associated with a ruptured disc is experienced. Depending on the extent and nature of the extrusion of this material against the neural tissue, some paralysis and loss of sensation may result. In a similar manner, a ligament that normally helps to support the vertebral column may push against the cord. This is particularly true in the elderly, in whom the ligamentum flavum is prone to buckle (Hockberger, 1988) and during neck extension may push against the cord and produce an injury, for instance, to the inner diameter of the cord—the so-called "central cord" syndrome (Duckworth, 1984). Consequently, the overriding concern with regard to injuries of the discs and ligaments is essentially the same as for fractures of the vertebrae, namely, will they cause injury to the spinal cord (or one of the peripheral nerves emerging from either side of the cord)?

Thus, a so-called "broken back" (this usually refers to a fracture of one or more vertebrae of the lumbar or thoracic region) or even a so-called "broken neck" (fracture of one or more cervical vertebrae), although serious in their own right, become much more severe if the bone is broken in such a way that the bone then injures the spinal cord (Figure F).

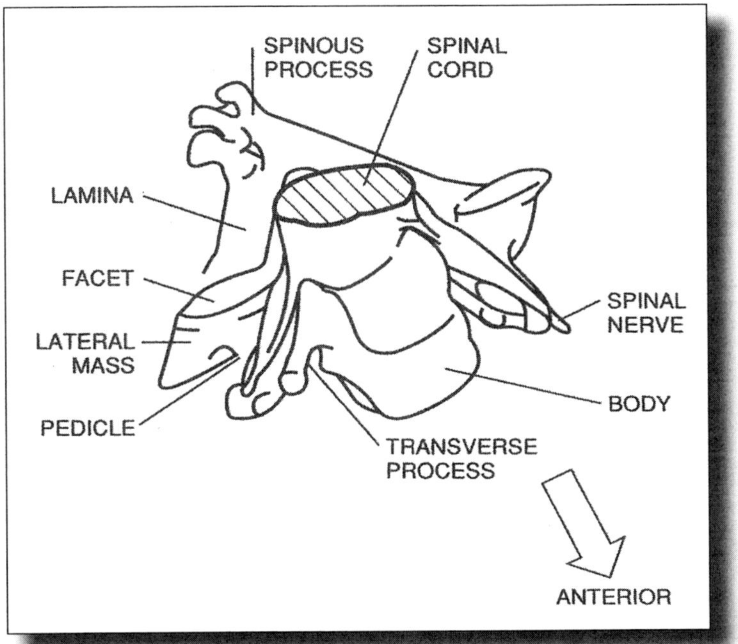

Figure E *Vertebra and the spinal cord. (From Pike, 1990, Figure 2-13. Published by SAE International. Used with permission.)*

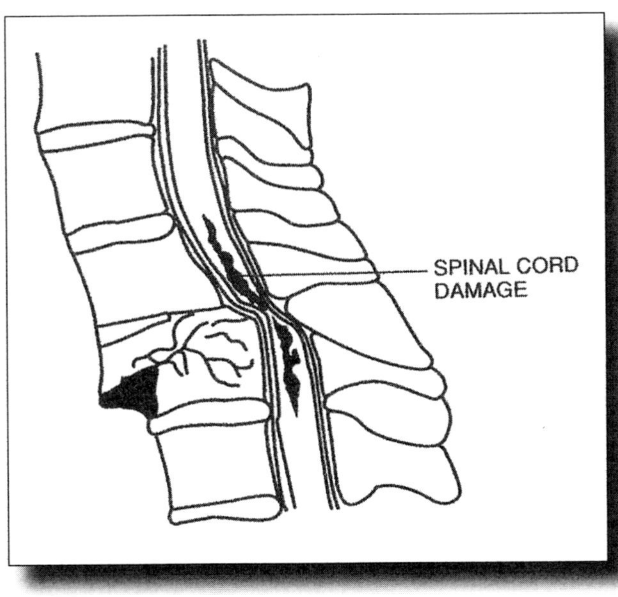

Figure F *Vertebral fracture with cord injury. (From Pike, 1990, Figure 2-17. Published by SAE International. Used with permission.)*

A complete disruption of the spinal cord, at any particular level, disrupts the functioning of the spinal cord at the site of transaction and distally, both sensory and motor. For example, if the spinal cord is severed at the level of T5, all spinal cord function handled by the cord below the T5 level will be lost. Similarly, if the spinal cord is severed at the T4 level, all of the spinal cord functions below the T4 level will be lost. Thus, a spinal cord transection at the T4 level would be equivalent to the loss due to a transection at the T5 level, plus the additional loss of the functioning of the cord segment from T5 to T4.

As a general guideline, the lumbar and sacral region of the spinal cord controls defecation, urination, and sexual function and contributes to ambulation (walking). The lumbar region controls the lower limbs and contributes to urination, the thoracic region controls the lower torso, and the cervical region controls the upper torso and upper extremities. Thus, transection of the spinal cord in the lumbar region generally would result in at least a partial loss of lower limb movement (paresis) and possibly a total loss of lower limb movement (paralysis or plegia).

Typically, the spinal cord per se terminates at the L1 level, and lumbar, sacral, and coccygeal nerve roots emerge from this point and continue down the spinal canal as the *cauda equina* (horse's tail). Generally, if the cord is transected above the L2 level, ambulation, even with a cane or brace, cannot be achieved. A spinal cord transection in the thoracic region would mean paralysis of the entire region below the waist. This also may include the region between the waist and mid-chest. (This loss still would be considered paraplegia because the paralysis affects the lower limbs but not the upper limbs.)

In general, spinal cord transection anywhere in the cervical region would result in quadriplegia (sometimes referred to as tetraplegia). Sometimes the term "quadraplegia" is used to indicate total paralysis of all four limbs, whereas the term "quadriparesis" is used to indicate partial loss of motion in all four limbs.

Unfortunately, cervical cord injuries, if in the upper cervical region, can have an additional deleterious effect. The phrenic nerve, which controls the respiratory diaphragm (and hence the ability to breathe), usually is a combination of nerve branches from the third, fourth, and fifth cervical levels. If the cord is severed above this level, the patient would be unable to breathe; hence, death would result. (An exception to this would be if the patient could be kept alive via artificial respiration until transported to a medical facility. The patient then could be helped to breathe, either by being ventilated mechanically or made to breathe "on his or her own" with the assistance of a phrenic nerve pacer [similar to a heart pacemaker].)

Note that in practice, most blunt trauma spinal injuries do not involve complete transection of the cord; thus, there may be motor and sensory losses at different levels. Furthermore, blunt trauma may not produce physical interruption of the cord, but rather may set in motion a biochemical cascade that results in tissue death within a few hours. Thus, there is the possibility of medical/surgical intervention to interfere with this process and improve the outcome. A number of systems have been developed for ranking and quantifying injury severity. A discussion of a number of these, including the AIS (abbreviated injury scale) and ISS (injury severity score), is presented in Pike, 1990. The American Spinal Injury Association (ASIA) impairment scale is available via the ASIA website at www.asia-spinalinjury.org.

Injuries of the spine in general are discussed in Pike, 1990, and injuries of the cervical spine are discussed in detail in Pike, 2002. The injuries above frequently are due to extremes of the various loadings and movements that otherwise are part of normal functioning of the spine. These include tension, compression, flexion, extension, lateral flexion (lateral bending), rotation, and combinations thereof.

The Head

The head may be conceptualized as consisting of three components: (1) the skin and other soft tissue covering the skull, (2) the skull (cranial and facial bones), and (3) within the skull, numerous blood vessels, three membranes (meninges), and the brain. Any of these components can be injured by a head impact. The brain and other intracranial contents also can be injured by abrupt head motion, even without significant head impact. Scalp injuries may be classified as superficial or deep, with deeper injuries being more significant with regard to bleeding, infection, and scarring. The bony structure of the head includes the cranium (frontal bone [forehead], temporal bone [above and surrounding the ear], parietal bone [rear side], and occipital bone [lower rear]) and the face (the maxilla [upper jaw bone], mandible [lower jaw bone], zygoma [cheek bone], and nasal bones) (Figure G). Within the skull, the brain is

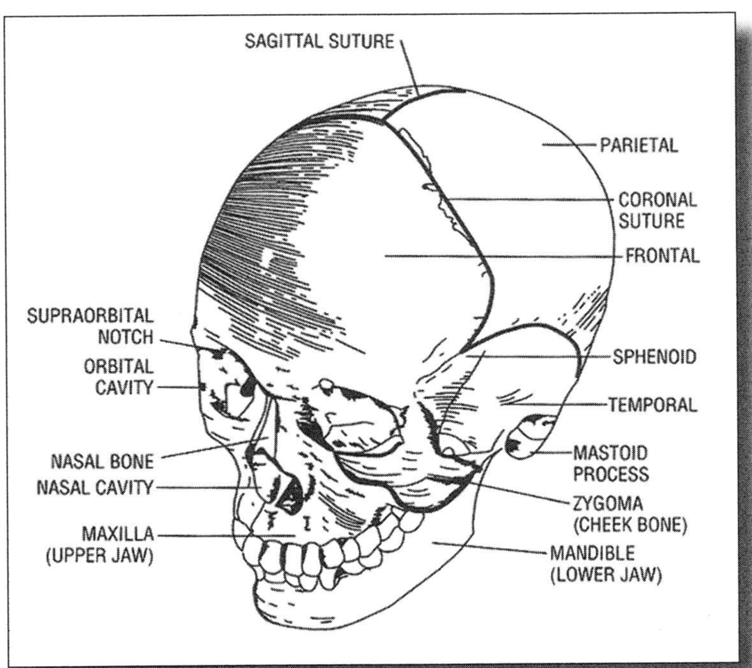

Figure G *Oblique view of the skull. (From Pike, 1990, Figure 2-3. Published by SAE International. Used with permission.)*

protected by three membranes. First, the outermost layer, the dura mater, provides a watertight covering and interface between the skull and the other meninges. Second, the filmy arachnoid layer serves as a conduit for blood vessels and as a container for cerebrospinal fluid. Third, the innermost membrane, the pia mater, provides an interface between the other membranes and the surface of the brain. Biomechanical studies have quantified forces of impacts that fracture the various bony regions, as well as accelerations and other parameters associated with concussion (Pike, 1990).

Much valuable information and insight regarding injury and injury mechanisms have been obtained from biomechanical studies of field data (i.e., vehicle crash databases, investigations of individual accidents, and combining accident investigation with medical record review), human volunteer tests, and crash dummy tests. These various sources enable biomechanics to study injury mechanisms and thereby to better understand how injuries occurred and how to mitigate future injuries. A well-established and widely used methodology (National Highway Traffic Safety Administration, 1999; Fricke, 1990) uses vehicle crush as a basis for determining impact severity, as measured by "delta v," an engineering term related to impact speed. As crash data recorders on passenger vehicles and trucks become more widespread and more uniform, field data will provide additional insight. Testing with various surrogates and, at the lower end of the severity spectrum, testing with human volunteers, provides quantitative data that can be compared to established reference values and limits to further refine tolerance values and to assess the likelihood of injury. In many countries, dummy crash testing is used to assess vehicle safety by comparing dummy test readings to mandated quantitative levels of performance. In the United States, the traditional parameters include the following: the acceleration-based Head Injury Criterion (HIC) of 1000, chest acceleration of 60 g's, chest compression of 3 inches and femur compressive loads of 2250 lb for frontal impacts (Pike, 1990), and 85-90 TTI (Thoracic Trauma Index) (an acceleration-based parameter) and 130 g's pelvic acceleration for side impacts (National Highway Traffic Safety Administration, 1990). More recently, the frontal requirements have been modified to reflect occupant age and size. Limits for neck flexion, extension, compression, tension, and shear were proposed by the National Highway Traffic Safety Administration in January 1997 and were adapted in a modified proposal using combined values (National Highway Traffic Safety Administration, 1999). Guideline values for other measurements (e.g., tibia force and ankle moments) also are being developed.

Chapter 1

א

Narratives

Chronologically, the first medical record is often prepared by the emergency medical services (EMS) personnel, who treat at the scene and may transport to a medical facility. The EMS care providers usually prepare a pre-hospital care report (PCR) (Bledsoe, 2003), which is often referred to as an EMS sheet or an EMS run sheet. The typical EMS run sheet provides information using both check boxes and narratives. The information includes patient identification, patient medical history, an assessment, and the treatment plan.

Usually at the top of the form, there is space for patient information, including name, address, and date of birth (DOB). There also is information regarding the EMS personnel, including the time and location of their first contact with the patient. In addition, this area of the form has space for the chief complaint (C/C), such as "MVA head injury" or "C/O neck pain" (complains of neck pain). Next is a section for the narrative, which describes the basics of the scenario, including subjective, objective, and assessment/management plan, findings, and causes of injury. If the report involves a motor vehicle accident (MVA), the narrative also may include some information about damage to the vehicle and occupant restraints. Note that the entries regarding restraints, such as seat belt usage, may be a repetition of what the patient tells the EMS, or they may be based on independent EMS observations (e.g., as part of patient extrication, EMS personnel cut the seatbelt, which was still holding the patient in place).

The narrative often is organized using one of two systems, so-called SOAP notes or CHART notes (Figures 1-1 and 1-2). SOAP (subjective, objective, assessment, and plan) and CHART (chief complaint, history, assessment, Rx, transport) can both serve as acronyms and will be discussed in more detail later. As is the case in Figure 1-1, each letter of the acronym may actually be written on the record as a section heading. Note on the left-hand side of the page, about halfway down, there is a handwritten letter "S," and five lines from the bottom of the page, the letter "O." Looking down on the page, three lines from the bottom, A and P are combined in the notation "A/P." Note that there is extensive use of abbreviations in the narrative as well as elsewhere in medical records, and the abbreviations that may be encountered are not entirely standardized. Some of the more common abbreviations and their definitions are provided in the Glossary of this book.

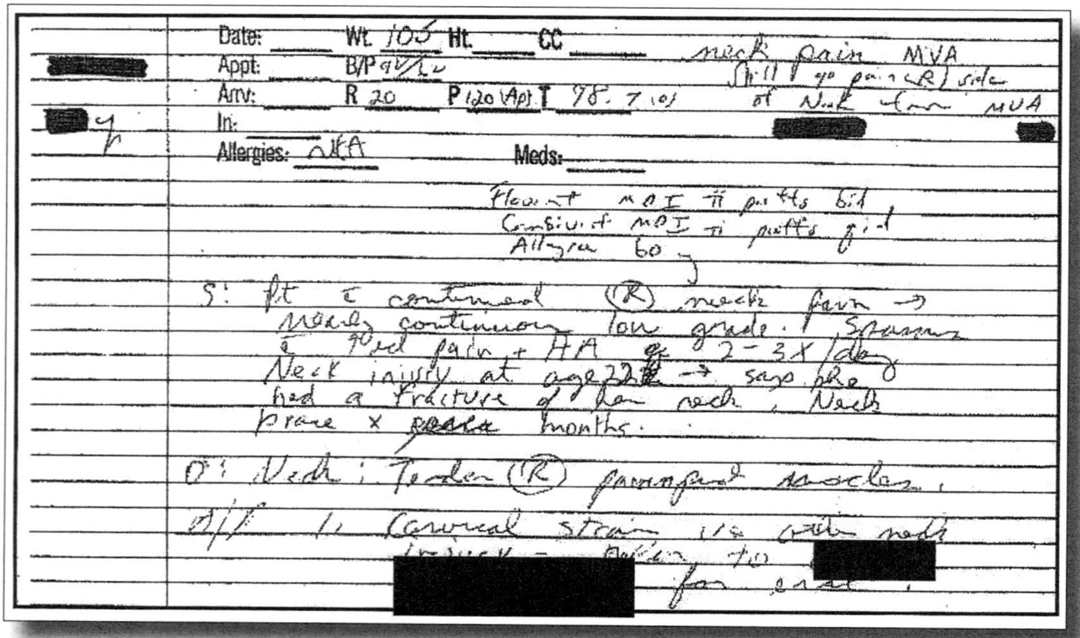
[redacted]
Date: ___ Wt. 105 Ht. ___ CC ___ neck pain MVA
Appt: ___ B/P 90/60 ___ Still ... pain (R) side of Neck ... MVA
Arrv: ___ R 20 P 120 (Ap) T 98.7 ...
[redacted] y/r
In: [redacted] [redacted]
Allergies: NKA Meds: ___

Flovent MDI ii puffs bid
Combivent MDI ii puffs qid
Allegra 60

S: Pt c continued (R) neck pain → nearly continuous low grade. Spasms c ... pain + HA ... 2-3X/day. Neck injury at age 22 → says she had a fracture of her neck. Neck brace x ~~...~~ months.

O: Neck: Tender (R) paraspinal muscles.

A/P 1. Cervical strain h/o cervical neck injury → ... to [redacted] for eval.

Figure 1-1 *SOAP notes.*

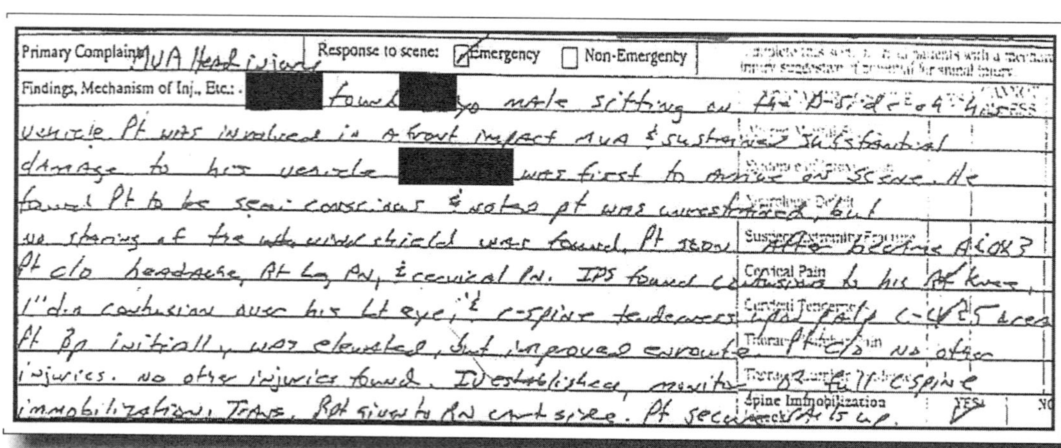
Primary Complaint: MVA Head injury | Response to scene: ☑ Emergency ☐ Non-Emergency

Findings, Mechanism of Inj., Etc.: [redacted] found [redacted] y/o male sitting on the D-side of his vehicle. Pt was involved in a front impact MVA c sustained substantial damage to his vehicle. [redacted] was first to arrive on scene. He found Pt to be semi conscious & notes pt was unrestrained, but no starring of the windshield was found. Pt ... A&Ox3 Pt c/o headache, Rt Leg, Arm, c cervical pain. IPS found contusion to his Rt knee, 1" dia contusion over his Lt eye, c c-spine tenderness upon palp C-4/C5 area. Pt Bp initially was elevated, but improved enroute. Pt c/o no other injuries. No other injuries found. IV established, monitor, full c-spine immobilization, Trans. Rpt given to RN and staff. Pt secured ...

Spine Immobilization YES

Figure 1-2 *Run CHART narrative.*

For now, perhaps it will suffice to mention that one of the most useful aspects of these systems is that they help to separate subjective (e.g., history of present illness, chief complaint) and objective (e.g., pulse rate, physical exam) entries. The following EMS run sheet excerpt, taken from the "Findings, Mechanism of Injury, Etc." section (Figure 1-2), can be read as follows:

> *__ found [a] __ year old male sitting on the d[driver]-side of his vehicle. Patient was involved in a front impact MVA and sustained substantial damage to his vehicle. __ was first to arrive on scene. He found patient to be semi-conscious and noted patient was unrestrained, but no starring [a star-burst glass fracture pattern, discussed in Case Study 11, Chapter 6, p. 71] of the windshield was found. Patient soon after became A&OX3 [alert and oriented times three, i.e., oriented with respect to person, place, and time, that is, they know who they are, where they are, and when it is]. Patient complained of headaches, right leg pain, and cervical pain. __ found contusions to his right knee, 1-inch diameter contusion over his left eye and c-spine [cervical spine] tenderness upon palpation [in the] C4/C5 area [mid-neck]. Patient's blood pressure initially was elevated, but improved en route. Patient complained of no other injuries. No other injuries found. IV [intravenous line] established, monitored oxygen, full c-spine immobilization. Transported. Report given to registered nurse curbside. Patient secured, rails up. [The rails on the side of the gurney were raised to limit patient sideways movement.]*

Beneath this are spaces to report various measures of patient function (e.g., blood pressure, lung sounds, pupil and skin condition). Under "Pupils," the entry PERL (pupils equal and reactive to light) is an indication of normal brain and cranial nerve function. Next a few lines are provided for "History/Medications/Allergies." These entries would be read as:

> *History of high blood pressure, diabetic. Meds [medication information not available]. Allergies: NKDA [no known drug allergy]. Next is a section that includes ECG [electrocardiogram] interpretation and the entry of NSR indicating "normal sinus rhythm" (heart beating normally), and under "Treatment–Airway Management/O2/Defib/Medication" reports if any treatment was needed to provide airway access and whether oxygen was provided, whether electrical defibrillation was provided, any intravenous lines, and any medication.*

The next record is a nursing flow sheet (Figure 1-3) from the emergency department and typically contains a chronologic record of treatment and observations. This particular sheet contains some rather untypical entries:

> *X-ray came to pick patient up. Patient acting as [if] passed out. Dr. ___ at bedside. Smelling salt applied. Patient willfully not breathing.*

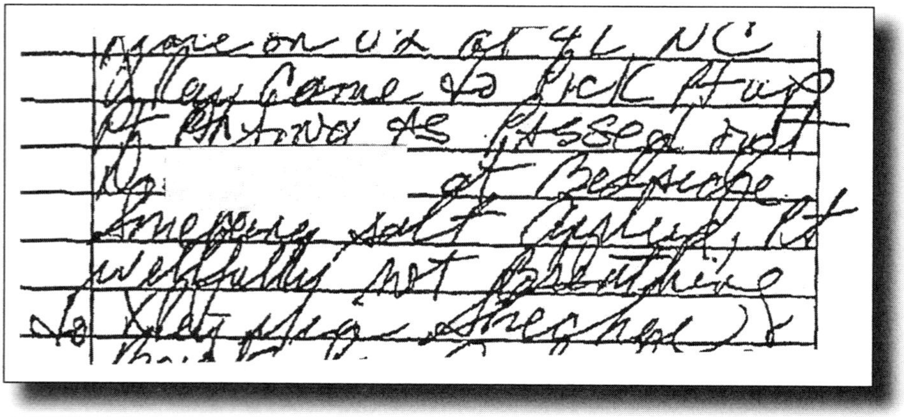

Figure 1-3 *Nursing flow sheet.*

Thus, for some reason, there appears to be an emotional/behavioral component to what is being observed, above and beyond whatever physical trauma the patient may have experienced.

The next record is a cervical spine x-ray (Figure 1-4), which states that the neck was non-tender and was "cleared clinically." By clearing the cervical spine, a clinician made the judgment call that any supportive cervical collars can be removed and that the patient can resume voluntary neck motion. In effect, this is opining that there is no significant risk of neurological injury from further voluntary neck motion (Pike, 2002).

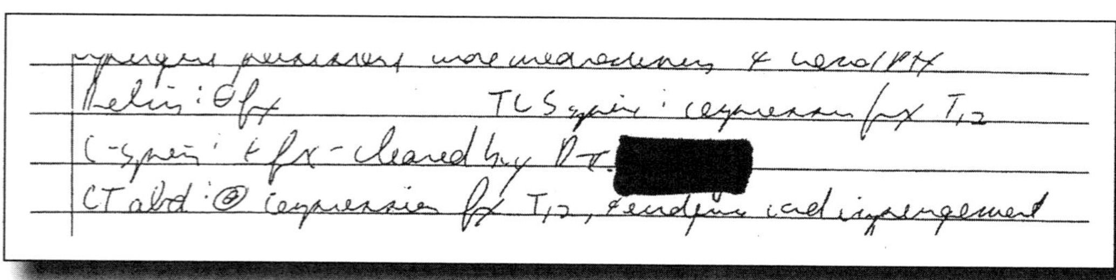

Figure 1-4 *Cervical spine cleared.*

This physician's progress record (Figure 1-5) reports on several imaging studies, including the cervical spine. The findings say that the pelvic region is negative for fracture and that the only finding in the thoracic lumbar and sacral region of the spine is a compression fracture at T12. The next notation, for the cervical spine, states "negative for fracture" and then states that it was "cleared by Dr. ____." The note continues by saying that a computed axial

Chest x-ray is unremarkable for air space disease or abnormal mediastinum. Neckwas nontender and she was cleared clinically.

Figure 1-5 *Physician progress record.*

tomography (CT) scan of the abdomen was positive for compression fracture of T12 (thereby confirming the finding on plain x-ray). However, there was no evidence of cord impingement, that is, although the bone of the twelfth thoracic vertebra was broken, there was no indication that a fragment of bone was pressing against the spinal cord.

This next record (Figure 1-6) states that the "oculocephalic not attempted." The oculocephalic refers to testing for brain stem integrity, which involves moving the head in a "no" and "yes" gesture and observing the resulting eye movement pattern (Plum, 1982). This was not attempted because at the time, the patient's cervical spine had not yet been "cleared." Thus, the judgment was made that whatever additional diagnostic information the oculocephalic test might have provided was not worth the associated risk of inducing spinal cord injury.

intensive care unit, he had anisocoria with the left and right pupil being 3.5 and 2 mm and nonreactive bilaterally. Oculocephalic not attempted, as the patient's C-spine had not been cleared at that time. There was some bloody discharge of the left ear, and a right hemotympanum was noted.

Figure 1-6 *Cervical spine not cleared.*

The next record is a death certificate (Figure 1-7). Note that the top line shows check boxes for six categories of death. In this instance, the assessment was made that this death was "accidental." Reporting this determination—that the death was accidental rather than natural, suicide, or homicide—is one of the primary functions of the certificate. Further information is provided: Item 28D, How Injury Occurred, states "one car accident," and under Item 30, Cause of Death, Item a states "Acute Cardiopulmonary Arrest." (Note that the instruction for this item specifically states, "Do not enter the mode of dying, such as cardiac or respiratory arrest.") In any event, Item b states this was due to "Fracture of the cervical spine, close head

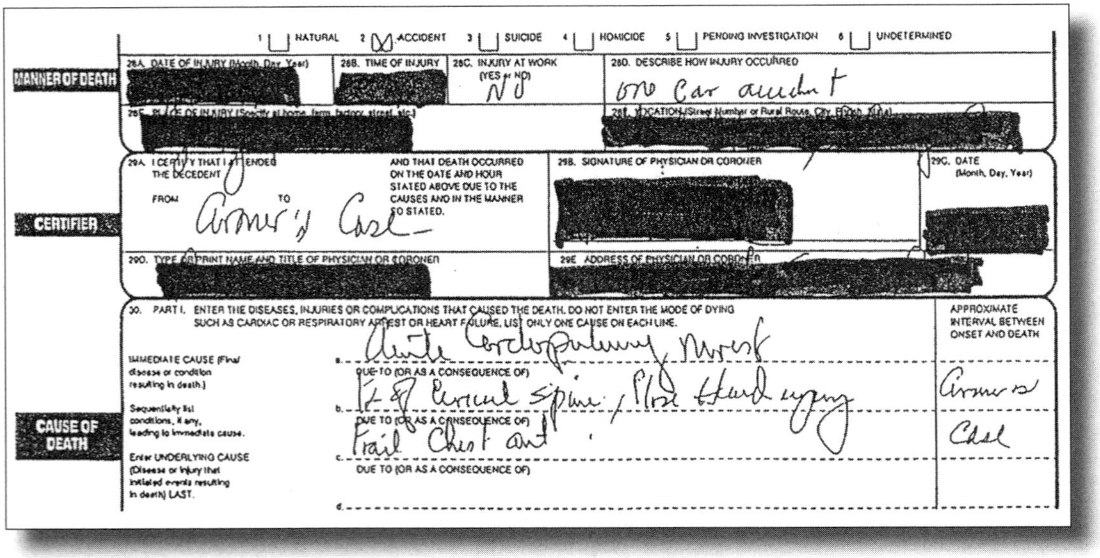

MANNER OF DEATH

1 ☐ NATURAL 2 ☒ ACCIDENT 3 ☐ SUICIDE 4 ☐ HOMICIDE 5 ☐ PENDING INVESTIGATION 6 ☐ UNDETERMINED

28A. DATE OF INJURY (Month, Day, Year)	28B. TIME OF INJURY	28C. INJURY AT WORK (YES or NO)	28D. DESCRIBE HOW INJURY OCCURRED
[redacted]	[redacted]	NO	one car accident

28E. PLACE OF INJURY (Specify at home, farm, factory, street, etc.) [redacted]

28F. LOCATION (Street Number or Rural Route, City, Borough, State) [redacted]

CERTIFIER

29A. I CERTIFY THAT I ATTENDED THE DECEDENT FROM Coroner's TO Case AND THAT DEATH OCCURRED ON THE DATE AND HOUR STATED ABOVE DUE TO THE CAUSES AND IN THE MANNER SO STATED.

29B. SIGNATURE OF PHYSICIAN OR CORONER [redacted]

29C. DATE (Month, Day, Year) [redacted]

29D. TYPE OR PRINT NAME AND TITLE OF PHYSICIAN OR CORONER [redacted]

29E. ADDRESS OF PHYSICIAN OR CORONER [redacted]

CAUSE OF DEATH

30. PART I. ENTER THE DISEASES, INJURIES OR COMPLICATIONS THAT CAUSED THE DEATH. DO NOT ENTER THE MODE OF DYING SUCH AS CARDIAC OR RESPIRATORY ARREST OR HEART FAILURE. LIST ONLY ONE CAUSE ON EACH LINE.

IMMEDIATE CAUSE (Final disease or condition resulting in death.)

Sequentially list conditions, if any, leading to immediate cause.

Enter UNDERLYING CAUSE (Disease or injury that initiated events resulting in death) LAST.

		APPROXIMATE INTERVAL BETWEEN ONSET AND DEATH
a.	Acute Cardiopulmonary Arrest	
b.	DUE TO (OR AS A CONSEQUENCE OF) Fx of Cervical spine / Close Head injury	Coroner's
c.	DUE TO (OR AS A CONSEQUENCE OF) Flail Chest ant	Case
d.	DUE TO (OR AS A CONSEQUENCE OF)	

Figure 1-7 *Death certificate.*

injury." (Note the instruction to list only one cause on each line.) Item c states the additional cause as "Flail chest, anterior," that is, sufficient rib fractures to produce a structural collapse of the front of the chest.

The next record, an autopsy report (Figure 1-8), gives the cause of death as a broken neck and does not provide additional detail such as vertebral level or the type of fracture. Also note that the deceased was an organ donor. Thus, at the time of the exam, the heart, cornea, and skin had already been harvested.

The next document also is a death certificate (Figure 1-9). Note that in this instance, the immediate cause of death was still being determined and was listed as "Pending Investigation."

In this next record (Figure 1-10), note that the cause of death originally was listed as under investigation. However, after the investigation was completed, the cause of death was changed from "Investigation Pending" to "blunt force craniocerebral injuries" secondary to a "motor vehicle accident" and that death was immediate. This report again has the six check boxes, and "accident" was selected. It also provides additional detail of the MVA, namely, the deceased was a driver who was stopped to make a left turn, was struck from behind, was sent into oncoming traffic, and thereby was involved in a head-on collision as well.

The next record (Figure 1-11) also reports on a death. Note that the bottom of this report summarizes that the probable cause of death as "multiple injuries." It was deemed an accident, and no autopsy was performed.

AUTOPSY REPORT

Case [redacted]

PATHOLOGICAL DIAGNOSES ON THE BODY

OF

[redacted]

1. Broken neck.
2. Postmortem absence of the heart, corneas and skin.

OPINION

It is our opinion that the decedent, came to his death as a result of broken neck, motor vehicle accident, passenger.

Figure 1-8 *Autopsy report.*

IMMEDIATE CAUSE (A) Pending Investigation

DUE TO (B)

Figure 1-9 *Cause of death being investigated.*

The next record (Figure 1-12) reflects the distinction between subjective and objective information. The record includes "Pupils fixed and dilated per EMS" and "Had hypertension and chest crepitus per EMS." Thus, we are informed that this is not based on the current clinician's observation; rather, they are recording information provided by an EMS person who was on the scene. Also note that this record reports that the patient was "Intubated on scene for GCS 3." The "GCS" refers to the Glasgow Coma Score, a numerical assessment of head injury severity. A value of 3 is indicative of very severe injury. As a result, the patient was intubated in the field. (An overview of the GCS is provided in Pike, 2002.)

PART I

INFORMATION AS IT APPEARS ON RECORD

107. DEATH WAS CAUSED BY ENTER ONLY ONE CAUSE PER LINE FOR A, B, C, AND D)

IMMEDIATE CAUSE (A) INVESTIGATION PENDING

(B)

(C)

DUE TO (D)

TIME INTERVAL BETWEEN ONSET AND DEATH

108. DEATH REPORTED TO CO[R] [X] YES REFERRAL NUMBER [redacted]

109. BIOPSY PERFORMED [ ] YES [X]

110. AUTOPSY PERFORMED [X] YES

111. USED IN DETERMINING C[AUSE] [X] YES

112. OTHER SIGNIFICANT CONDITIONS CONTRIBUTING TO DEATH BUT NOT RELATED TO CAUSE GIVEN IN 107

113. WAS OPERATION PERFORMED FOR ANY CONDITION IN ITEM 107 or 112? IF YES, LIST TYPE OF OPERATION AND DATE.

119. MANNER OF DEATH [ ] NATURAL [ ] SUICIDE [ ] HOMICIDE [ ] ACCIDENT [X] PENDING INVESTIGATION [ ] COULD NOT BE DETERMINED

120. INJURY AT WORK [ ] YES [ ] NO

121. INJURY DATE—MM / DD / CCYY

122. HOUR

123. PLACE OF INJURY

124. DESCRIBE HOW INJURY OCCURED (EVENTS WHICH RESULTED IN INJURY)

125. LOCATION (STREET AND NUMBER OR LOCATION AND CITY AND ZIP CODE)

PART II

INFORMATION AS IT SHOULD APPEAR

107. DEATH WAS CAUSED BY ENTER ONLY ONE CAUSE PER LINE FOR A, B, C, AND D)

IMMEDIATE CAUSE (A) BLUNT FORCE CRANIOCEREBRAL INJURIES — IMM

(B) MOTOR VEHICLE ACCIDENT — IMM

(C) –

DUE TO (D) –

TIME INTERVAL BETWEEN ONSET AND DEATH

108. DEATH REPORTED TO C[OR] [X] YES REFERRAL NUMBER [redacted]

109. BIOPSY PERFORMED [ ] YES [X]

110. AUTOPSY PERFORMED [X] YES

111. USED IN DETERMINING [X] YES

112. OTHER SIGNIFICANT CONDITIONS CONTRIBUTING TO DEATH BUT NOT RELATED TO CAUSE GIVEN IN 107

113. WAS OPERATION PERFORMED FOR ANY CONDITION IN ITEM 107 or 112? IF YES, LIST TYPE OF OPERATION AND DATE.

119. MANNER OF DEATH [ ] NATURAL [ ] SUICIDE [ ] HOMICIDE [X] ACCIDENT [ ] PENDING INVESTIGATION [ ] COULD NOT BE DETERMINED

120. INJURY AT WORK [ ] YES [X] NO

121. INJURY DATE—MM / DD / CCYY [redacted]

122. HOUR [redacted]

123. PLACE OF INJURY [redacted]

124. DESCRIBE HOW INJURY OCCURRED (EVENTS WHICH RESULTED IN INJURY)

Driver, stopped to make left turn, struck from beh
and sent into oncoming traffic. Head on collision.

Figure 1-10 *Investigation follow-up.*

PROBABLE CAUSE OF DEATH	MANNER OF DEATH		DISPOSITION OF CASE
Multiple Injuries	(Check one only)		1. Not a medical examiner case ☐
	☑ Accident	☐ Natural	
	☐ Suicide	☐ Unknown	2. Autopsy ☐ Yes ~~☐ No~~
	☐ Homicide	☐ Pending	Pathologist [redacted]

Figure 1-11 *Multiple injuries noted.*

ATTENDING NOTE mercy flight from scene of MVC — unknown restraint ⊖ air bag. Front end damage to vehicle. Intubated on scene for GCS 3. Pupils fixed & dilated per EMS. Had hypotension and chest crepitus per EMS.
PMH - unknown.

Figure 1-12 *Note cites information provided by EMS.*

The next record (Figure 1-13) illustrates the importance of imaging the entire cervical spine before the patient is "cleared." Clearing refers to the decision, made by a health care provider (generally a physician) relatively early in a patient's care, that the nature and extent of cervical injury has been adequately assessed to determine that the patient will not be at any appreciable risk of additional cervical injury from ordinary neck movement. Two important consequences of a cleared spine are that precautions such as neck immobilizations (e.g., neck brace) may be discontinued, and the patient may move his or her neck as required to have additional images taken (Pike, 2002). Note that the image being discussed only visualizes cervical vertebrae C1 through C6. The radiologist is very clear about requiring visualization of the rest of the cervical spine (i.e., the C6/C7 disc, C7, and the C7/T1 disc) perhaps from more than one direction, when he writes, "Additional views are required to clear the cervical spine radiographically."

IMPRESSIONS

1. LIMITED STUDY OF THE CERVICAL SPINE WITH ONLY A LATERAL VIEW. C1 THROUGH C6 APPEAR NORMAL ON THE LATERAL VIEW. ADDITIONAL VIEWS ARE REQUIRED TO CLEAR THE CERVICAL SPINE RADIOGRAPHICALLY.

Figure 1-13 *Cervical spine not cleared.*

Chapter 2

ℵ

Imaging

Background

Diagnostic imaging may be defined briefly as "a visual presentation of patient information, which is used to help make a diagnosis" (Pike, 2002). The image usually is based on information from the inside of the body; hence, much diagnostic imaging is based on information conveyed through non-visible energy, such as x-rays, which can pass through the skin and other structures (Figure 2-1). Furthermore, some diagnostic images convey information about the body's function, rather than structure, and this information might not be visibly discernible even if the body were surgically "opened." This discussion will focus on three imaging modalities: plain film radiography (x-rays), computer tomography (CT), and magnetic resonance imaging (MRI or simply MR).

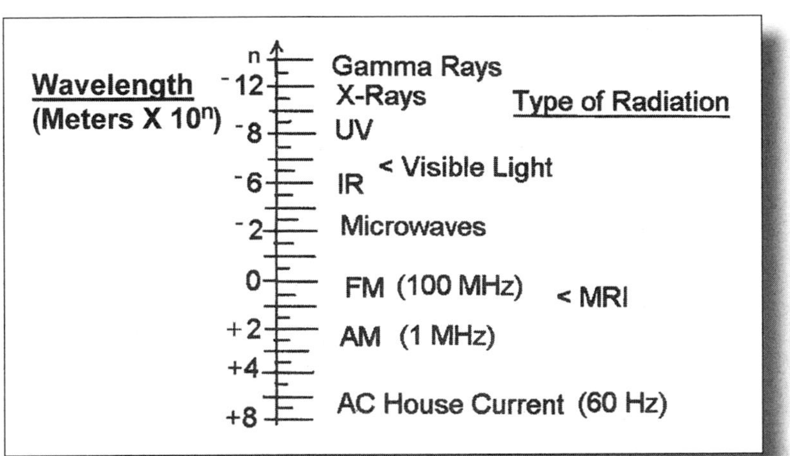

Figure 2-1 *Wavelengths and frequencies. (Pike, 2002, Figure 2-2. Published by SAE International. Used with permission.)*

One common way of classifying the energy used for imaging is by its wavelength. There is a continuum of wavelengths, and within this continuum, various groupings (bands) of wavelengths that have similar properties have been identified. Wavelengths are specified in units of length, and the basic unit is the meter (approximately 39.4 inches). A unique frequency is associated with radiation of each wavelength; thus, radiation may alternatively be characterized and grouped by its characteristic frequency. The basic unit of frequency is the hertz (one cycle per second). Visible light has a much higher frequency than radio waves, and x-rays in turn have much higher frequencies than visible light (Figure 2-1).

A more radiodense structure appears lighter on a standard plain film radiographic image, and a less radiodense structure appears darker. This is similar to a negative of regular photographic film—objects that are relatively dark when a print is made appear relatively light on the negative, and vice versa. Thus, the vertebrae appear rather white (Figure 2-2), they have a high physical density, and they produce an x-ray image of low optical density. Difficulties in interpreting radiographs may be due to a variety of sources, including the presence of congenital or developmental abnormalities, superposition of other structures, variations of normal anatomy (e.g., obesity), and limitations of the imaging system. For instance, a number of conditions may give the appearance of fractures even though no fracture is present. An example of one of these, the so-called "Mach effect," is provided in Pike, 2002.

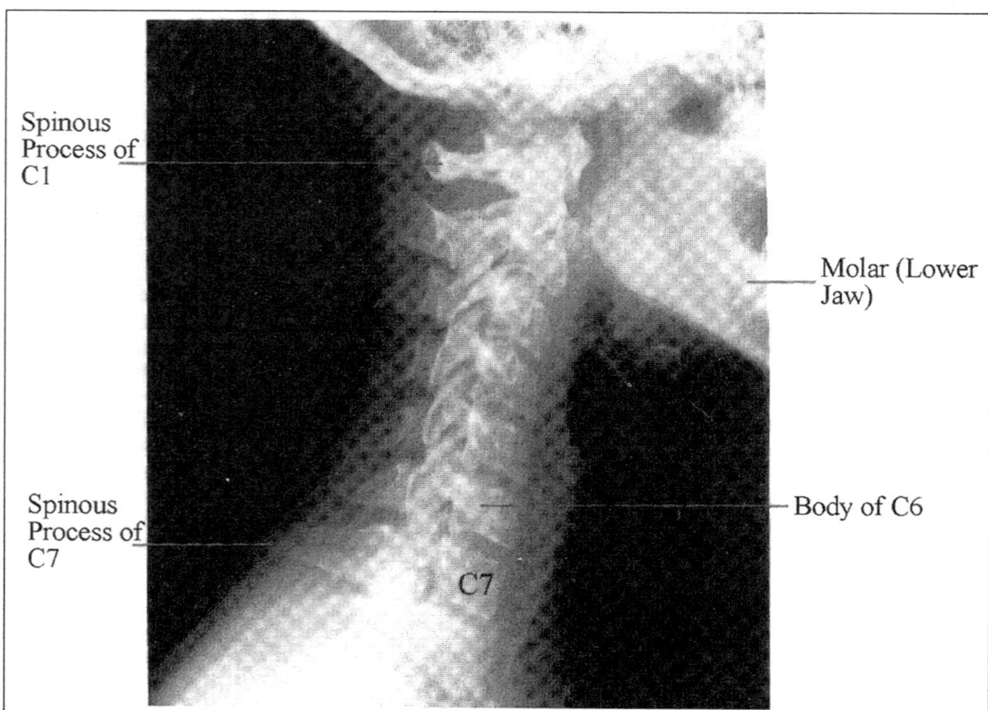

Figure 2-2 *An x-ray of the cervical spine. (Pike, 2002, Figure 2-1. Published by SAE International. Used with permission.)*

An important innovation in x-ray imaging permits viewing a "slice" of the subject rather than the traditional view of the entire depth. Such an image is called a tomogram. Tomograms were invented about 50 years ago but were generated by mechanical means until the early 1970s, when the first tomogram was produced using computer technology. A more recent innovation (1980s) called magnetic resonance imaging (MRI or MR) does not even use x-rays. Indeed, one of its most desirable features is thought to be that it uses only lower-frequency, lower-energy radiation. The radio waves used in diagnostic MR do not produce an image by passing through the body region of interest but rather, in the presence of a strong magnetic field, are caused to be emitted by the region of interest. This technique for obtaining images is referred to as emission, as compared to transmission for x-ray-generated images. This section will emphasize reports of x-rays, CTs, and MRIs.

Discussion

As is typical of a radiology report, the first part of the narrative of this document (Figures 2-3 and 2-4) is descriptive, that is, the body region(s) imaged, the imaging procedure used, and the results observed. The second part of the report is subjective (impressions). This radiology report provides an example of how the medical records may be interpreted to address impact direction, fracture mechanism, and pre-existing conditions.

The top-most cervical vertebra, C1, is located just below the base of the skull (Figure D in the Background section of this book and Figure 2-2) and provides a transition between the skull and the cervical spine. The second cervical vertebra, C2, is adjacent to and below C1. The odontoid process is the vertical post-like projection of C2, around which the ring-shaped C1 rotates. Thus, the "anterior subluxation of C1 and the odontoid with respect to the body of C2 and anterior angulation" is saying that the upper part of C2 has been broken and, along with C1, has been translated forward with respect to the lower part of C2. The dens (odontoid) fragment also has been tilted forward. Thus, we can conclude that this injury most likely was due to neck flexion (i.e., the neck bending forward) (Anderson, 1989; Pike, 2002). There also is left lateral subluxation of the odontoid fracture, so the fracture-producing motion may not have been pure flexion, but rather may have had a component of head movement toward the left as well. Thus, we would expect the vehicle impact to be frontal, but to the left of center, perhaps 10 or 11 o'clock. (A more detailed discussion of odontoid fractures, as well as other neck injuries, may be found in Pike, 2002.) If we also had vehicle deformation data, we could check for consistency between these two data sources.

The report also indicates that there is mild anterolisthesis of C4 on C5. Because C4 is directly above C5, this is saying that the middle part of the neck has moved forward with respect to the lower part. In conjunction with the previous information, this leads us to believe that the forward component of the crash was fairly significant—not only did it produce vertebral fracture and displacement in the upper neck, but it also displaced two vertebrae in the middle of the neck. Furthermore, the report states that there are widespread degenerative changes throughout the remainder of the cervical spine. The patient is a post-menopausal female; thus, spinal degeneration would not be completely surprising. However, in this instance, there was a combination of normal degenerative changes as well as degenerative changes following a

```
IMAGING DEPARTMENT

   NAME:                                  COMPLETED:
    MRN:                DOB:                PRINTED:
VISIT #:                                     ADM MD:
  ACC #:               SITE:                 ORD MD:
   EXAM : C-SPINE LIMITED                  /ER/STAT

   ICD-9 CODE(S): 723.1
   CPT-4 CODE(S): 72040
-------------------------------------------------------------------------
     CERVICAL SPINE LIMITED

     AP, AP odontoid and lateral views show fracture of the base of the
     odontoid with anterior subluxation of C1 and the odontoid with
     respect to the body of C2 and anterior angulation.  The anterior
     subluxation is approximately 10 mm.  The AP odontoid view shows left
     lateral subluxation of the odontoid fracture fragment and the C1 ring
     to the left relative to the right lateral mass of C2 of approximately
     15 mm.  A portion of the left lateral mass of C2 appears to be
     fractured and displaced as well.  There is mild anterolisthesis of C4
     on 5.  There are degenerative changes throughout the remainder of the
     cervical spine.  There is a probable left apical pneumothorax.  The
     films are reviewed with Dr.                  A CT scan is to be
     performed.

     IMPRESSION:

     Displaced fracture of the base of the odontoid with anterior and left
     lateral subluxation of the C1 odontoid fracture complex.

     Probable fractures of the left lateral mass of C2.

     Mild anterolisthesis of C4 on 5.

     Probable left apical pneumothorax.


     Dictated by:                          M.D.
```

Figure 2-3 *Radiology report, cervical spine limited series.*

previous trauma. Thus, we know that before this motor vehicle accident (MVA), the subject's spine had undergone gradual changes and over a period of time had become less flexible and more fragile (Boden, 1991; Pike, 1989). Although the records presented do not specifically describe the extent of the degenerative changes at the injury site, one interpretation would be that the bone structure was frail and either fractured more extensively than a healthier spine would have fractured or, indeed, a healthier spine may not have fractured at all in the exact same MVA. Note also that the Impression section of the radiology report does not include all of the radiologist's observations.

AP, AP odontoid and lateral views show fracture of the base of the odontoid with anterior subluxation of C1 and the odontoid with respect to the body of C2 and anterior angulation. The anterior subluxation is approximately 10 mm. The AP odontoid view shows left lateral subluxation of the odontoid fracture fragment and the C1 ring to the left relative to the right lateral mass of C2 of approximately 15 mm. A portion of the left lateral mass of C2 appears to be fractured and displaced as well. There is mild anterolisthesis of C4 on 5. There are degenerative changes throughout the remainder of the cervical spine. There is a probable left apical pneumothorax. The films are reviewed with Dr. A CT scan is to be performed.

IMPRESSION:

Displaced fracture of the base of the odontoid with anterior and left lateral subluxation of the C1 odontoid fracture complex.

Probable fractures of the left lateral mass of C2.

Mild anterolisthesis of C4 on 5.

Figure 2-4 *Radiology report, cervical spine limited series—close-up.*

This next radiology report (Figures 2-5 and 2-6) shows the potential importance of previous medical history and pre-existing conditions. The report indicates there is a (pre-existing) fusion of C5 through C7. Thus, three vertebrae in the lower part of the neck had been fused previously (i.e., mechanically joined together), using screws and metal plates and inserting grafted bone as "shims" between adjacent vertebrae. (Other records, not shown here, provide additional background information. There was a fracture at one of the sites where a screw was inserted into the bone to hold the metal plate in place and to help join the fused vertebrae together. Sometime after this, the person was in an MVA, when the neck was fractured again. However, this time, the bone graft was fractured.)

Referring to the radiology report, we see that although the radiologist did not know the background, he observed the grafting hardware and noticed a fracture "through the fusion mass at approximately the level of C7." Because three vertebrae were fused, it would be expected that if a fracture occurred, it would occur at one or both of the boundaries, and this is what happened. In contrast, a healthy spine, which could distribute the bending moment between more of the vertebrae, might have not fractured at all.

The next study uses two documents (Figures 2-7 and 2-8) regarding the same individual to show how different records for the same person may provide different background information or may provide different perspectives on the same basic information. The first document, an excerpt from a typed "History and Physical," states that "she had no prior history of neck problems for at least the past 30 years." Thus, one might interpret this to mean that the patient had a very healthy neck, without even some of the more common aches and pains, since her 20s. (The patient's current age has been blocked out, but she is in her 50s.) The second document,

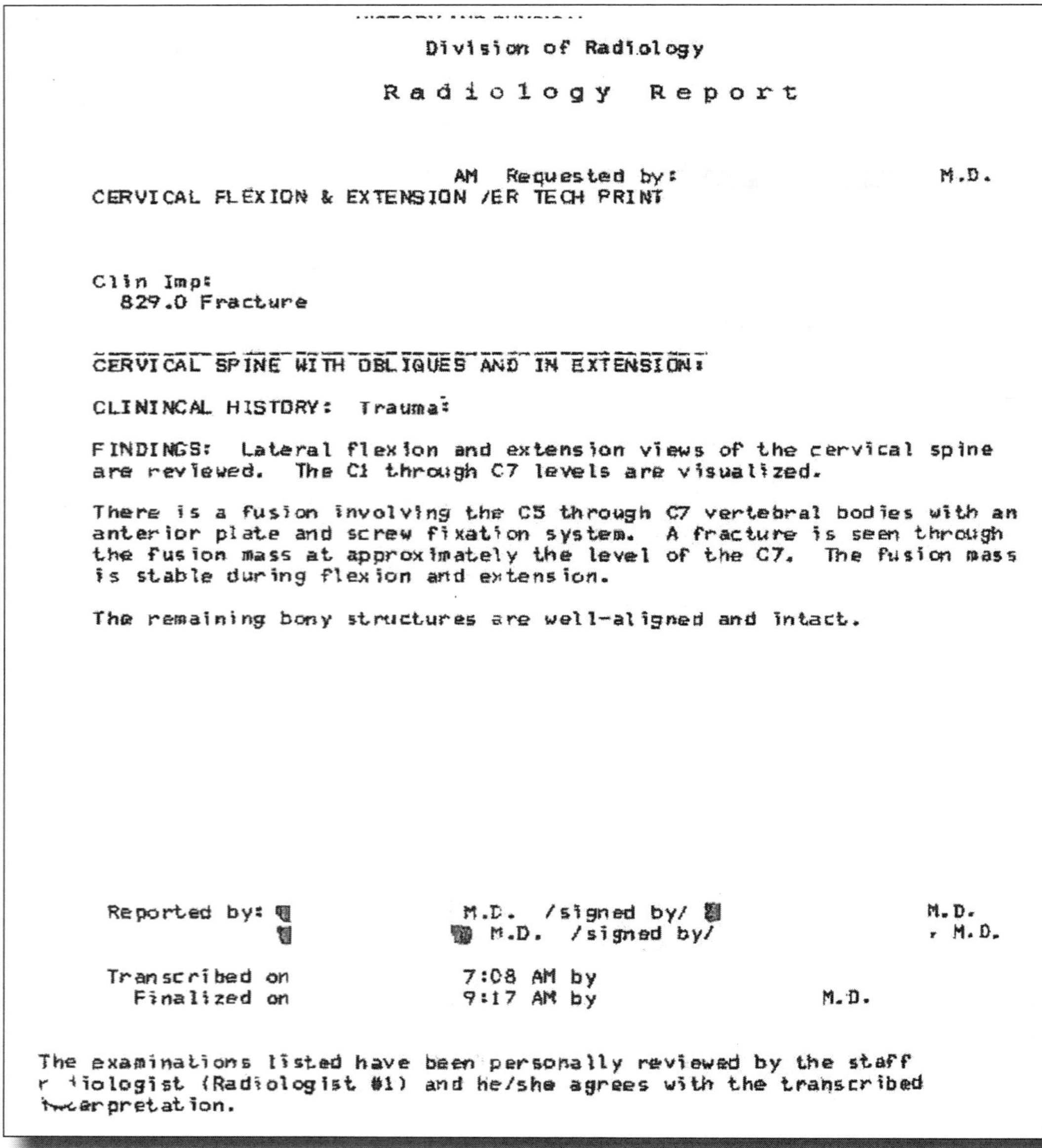

Division of Radiology

Radiology Report

AM Requested by: M.D.
CERVICAL FLEXION & EXTENSION /ER TECH PRINT

Clin Imp:
829.0 Fracture

CERVICAL SPINE WITH OBLIQUES AND IN EXTENSION:

CLININCAL HISTORY: Trauma:

FINDINGS: Lateral flexion and extension views of the cervical spine are reviewed. The C1 through C7 levels are visualized.

There is a fusion involving the C5 through C7 vertebral bodies with an anterior plate and screw fixation system. A fracture is seen through the fusion mass at approximately the level of the C7. The fusion mass is stable during flexion and extension.

The remaining bony structures are well-aligned and intact.

Reported by: [redacted] M.D. /signed by/ [redacted] M.D.
[redacted] [redacted] M.D. /signed by/ [redacted], M.D.

Transcribed on 7:08 AM by
Finalized on 9:17 AM by M.D.

The examinations listed have been personally reviewed by the staff r iiologist (Radiologist #1) and he/she agrees with the transcribed i~erpretation.

Figure 2-5 *Radiology report on flexion and extension views.*

excerpted from a handwritten office note by a different clinician, describes the same history a bit differently: "Neck injury at age 22 → says she had a fracture of her neck."

Thus, from the first statement, one may conclude that any current complaint is completely of recent origin, whereas based on the second record, one might suspect a long-term post-fracture

CLININCAL HISTORY: Trauma:

FINDINGS: Lateral flexion and extension views of the cervical spine are reviewed. The C1 through C7 levels are visualized.

There is a fusion involving the C5 through C7 vertebral bodies with an anterior plate and screw fixation system. A fracture is seen through the fusion mass at approximately the level of the C7. The fusion mass is stable during flexion and extension.

The remaining bony structures are well-aligned and intact.

Figure 2-6 *Radiology report on flexion and extension views—close-up.*

HISTORY AND PHYSICAL

PATIENT:

DATE OF BIRTH:

CHIEF COMPLAINT: Cervical pain, right arm pain, right hand tingling, headache, right shoulder weakness.

PRESENT ILLNESS: Patient explained she was in an automobile accident (date estimated). She was driving and exiting on when she was rear ended by a truck. She explained she had no prior history of neck problems for at least the past 30 years.

Figure 2-7 *History I.*

S: pt c̄ continued (R) neck pain → nearly continuous low grade. Spasms c̄ ↑'ed pain + HA @ 2-3X/day
Neck injury at age 22 → says she had a fracture of her neck. Neck brace X months

Figure 2-8 *History II.*

degenerative process to be at work. Note that both statements essentially are accurate. Inter-record differences may reflect diverse factors such as the following:

1. How a patient is feeling about himself or herself, his or her past, and so forth at the time the different histories are being taken
2. Different perspectives on the part of different history takers
3. The patient's recollection at a particular point in time (i.e., people will remember different aspects of their past at different times)
4. The patient's focus at the time of a given history (e.g., if concerned about major pain, he or she may not mention minor pain)
5. Possible malingering

Regardless of the reason for the differences between the patient histories, the bottom line is that there may be some advantage to reviewing as many records as possible.

The next record is from a cervical spine series (Figure 2-9). In this instance, the radiologist addresses the time of occurrence of the conditions he observed, namely, his impression states "degenerative arthritis" changes. He has ascribed the lesions he has observed (e.g., disc space narrowing and spurring) to degenerative arthritis, a pre-existing and gradually evolving condition. In effect, this rules out that any of these lesions are attributable to recent trauma.

CERVICAL SPINE SERIES:

AP, lateral and oblique views of the cervical spine show straightening and slight reversal of the normal cervical lordosis. There is moderate intervertebral disk space narrowing at C5-6. The remainder of the intervertebral disk spaces are maintained. Moderate end plate osteophytic spurring is noted at C4 through C6 with mild spurring at C7. There is moderate neural foraminal narrowing on the right at C4-5 and C5-6. Mild neural foraminal narrowing on the left at C4-5 and C5-6. Neural foramina are otherwise normal caliber and the posterior elements are aligned.

IMPRESSION: DEGENERATIVE ARTHRITIC CHANGE C4-5 AND C5-6 BUT GREATEST AT THE C5-6 LEVEL.

Figure 2-9 *Cervical spine series.*

The next document is taken from a radiology consultation report (Figure 2-10). As detailed in the anatomy discussion in the Background section of this book, a healthy spine, when viewed from the side, will have a curve roughly corresponding to the cervical region. This curvature typically is convex forward and is referred to as cervical lordosis. If, at the time of imaging, the curve is relatively slight or is absent, it is often referred to as "cervical straightening." In

IMPRESSION:
Diminished lordosis, probably related to positioning or spasm. No acute cervical trauma.

Figure 2-10 *Cervical lordosis I.*

general, abnormalities may be classified as developmental, degenerative, traumatic, neoplastic, or post-operative (Daffner, 1993). More specifically, the cervical straightening may be attributable to one or more of several factors, including normal human variability, the patient's neck position at the time of imaging, and neck muscle spasm (Harris, 1993). (Indeed, if the chin is voluntarily tucked down close to the chest, it has the effect of straightening the cervical spine. Also, if the patient is wearing a cervical collar, it has the effect of straightening the cervical spine.) Note that in this instance, the radiologist's Impression section of the report addresses the causes of the diminished lordosis (straightening) as "positioning or spasm."

This second example discussing cervical lordosis (Figure 2-11) includes five views of the cervical spine (AP [anterior-posterior], lateral, two obliques, and open mouth). (These views are discussed in detail in Pike, 2002.) The radiologist describes his findings as a "straightened normal lordotic curvature," which may be indicating curvature that although diminished from average is still in the normal range. The radiologist reports levoscoliosis (a curvature to the left). With regard to causation, the radiologist says "may or may not be positional in nature."

AP, lateral, oblique and open mouth views of the cervical spine (five views revealed a straightened normal lordotic curvature and levoscoliosis which may or may not be positional in nature Osteoporosis is seen. No evidence for acute fracture or dislocation

Figure 2-11 *Cervical lordosis II.*

This next record is a radiology consultation report (Figure 2-12). Note that Item 2 under the Impression section refers to the D1, D2, and D3 vertebrae. This appears to be only a typographical error or a transcription error. The thoracic vertebrae sometimes are referred to as the dorsal vertebrae; thus, a prefix "D" conceivably might be used to specify the thoracic vertebrae as D1, D2, and so forth. However, in this particular instance, the image is identified

2. There is cortical abnormality to the anterior aspect of the superior endplates of D1, D2, and to a lesser extent, D3 with more of a Schmorl's node impression on the anterior aspect of the superior endplate of D3. The findings at D1 through D3 are suggestive of a deceleration type injury. As there is no acute bone marrow edema, the finding is older than 3 months.

Figure 2-12 *Deceleration injury.*

as a cervical MRI. Therefore, the assumption will be made that these are referring to C1, C2, and C3—the first, second, and third cervical vertebrae. (These designations are discussed in the Background section of this book.) Note that the radiologist's impression addresses the injury mechanism and timing as well; he classifies it as "a deceleration type injury" and states that it occurred more than three months before the imaging. That is, the radiologist did not observe the bone marrow edema that such an injury would be expected to produce. Furthermore, because such edema usually resolves within a number of months, he opined that the injury was more than three months old. Thus, if the MVA occurred within three months of the imaging, this in effect would be saying that the lesions were not caused by the subject MVA.) Note that the radiologist also identifies a Schmorl's node at C3; this refers to a prolapse of material from the C2/C3 disc into the upper endplate of C3.

In this next record (Figure 2-13), the radiologist indicates that although he considers it unlikely that the lucent line he found on the radiograph represents a fracture, he does not feel that he has enough information to rule out a fracture. Therefore, he recommended additional imaging (CT) to be able to state his impression with a higher degree of confidence.

IMPRESSION:
Lucent line through the superior facet of C7. Fracture cannot be entirely excluded and CT scanning is recommended. This was discussed by telephone with ________ .

Figure 2-13 *CT recommended.*

This emergency room (ER) record (Figure 2-14), under the heading Diagnostics, reports that x-rays of the cervical spine show what appear to the clinician to be two fractures, the first transverse through what would be the area of C5, and the second oblique through what would be in the area of C7. These vertebral bodies have been removed and fused with an iliac graft,

X-rays of the cervical spine show what appear to me to be two fractures, the first transverse through what would be the area of C5 and the second oblique through what would be in the area of C7. These vertebral bodies have been removed and fused with an iliac graft and there is a plate on the anterior border of this cervical vertebral bodies.

Figure 2-14 *New injury at a previous injury site.*

and there is a plate on the anterior border of these cervical vertebral bodies. This report discusses injury at the site of a previous surgery, that is, several vertebrae were surgically fused together. This involves removing the discs and using bone harvested from the hip bone to act as a wedge. The wording in the Diagnostics section reflects that these are no longer vertebrae per se but rather refers to the fracture location as "what would be the area of C5" (if C5 were still intact). Note that before the current injury, there was a fusion of C5 through C7. This would create a very stiff and inflexible structure; thus, the spine would be prone to injury at the boundary between the fusion and the more flexible, non-fused spine, namely, at C4/C5 and at C7/T1.

The next record, an MRI of the lumbar region (Figure 2-15), includes a discussion of previous imaging. More specifically, comparison of imaging before and after the spinal fusion indicates a disc protrusion post-surgery that was not present pre-surgery. In this instance, the disc protrusion apparently occurred during or after the spinal surgery and not from the original trauma. (We do not know if there was some other possible causative event, in addition to the surgery, that occurred between the two imaging times.)

IMPRESSION:

Postop changes are seen to involve the lumbar spine with a posterior fusion from L2 to L4. Incidental small subligamentous disc protrusion /herniation at the L2-L3 level extending to the right which was not present on the preoperative study.

Figure 2-15 *Disc protrusion.*

The next record is a report of a bone scan (Figures 2-16 through 2-18), conducted to rule out pelvic injury and occult fractures in the lateral portion of ribs T7 through T10. The bone scan is different from the other imaging modalities discussed so far, in that rather than displaying a representation of anatomical structure that can be examined for irregularities, the bone scan provides an image that displays bone metabolism. The bone scan is not very

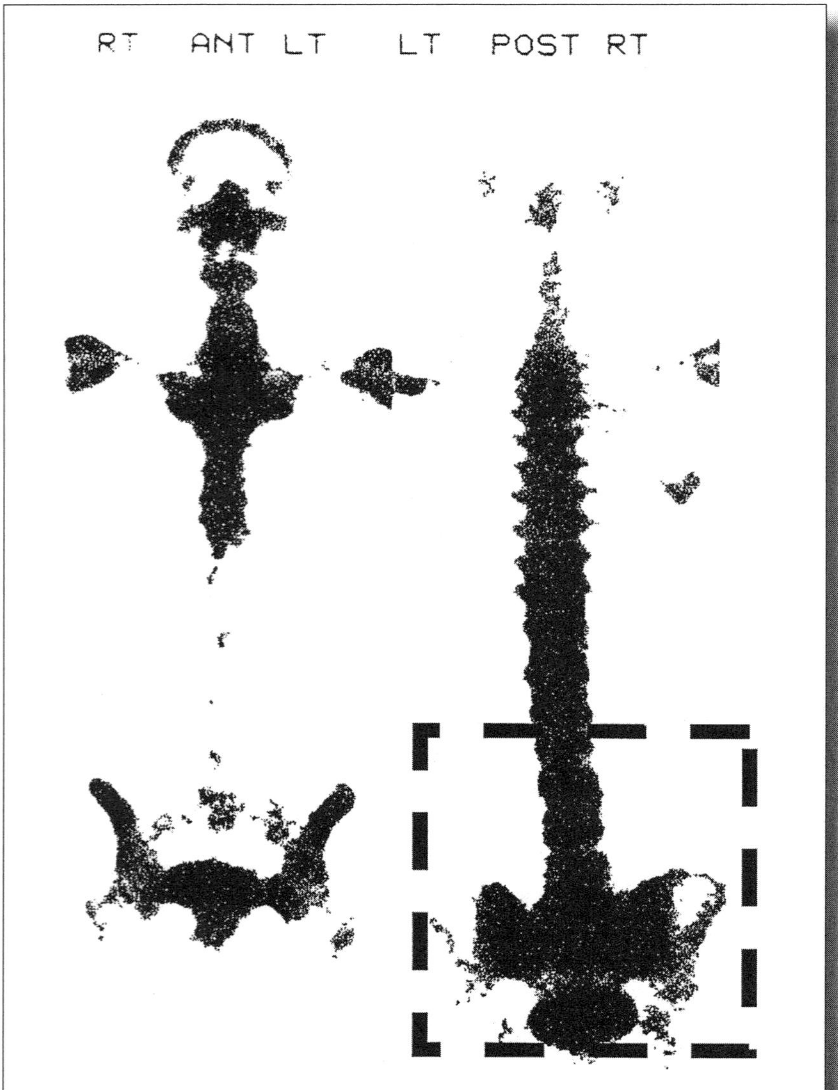

Figure 2-16 *Bone scan.*

good at indicating "what's wrong," but rather is useful to determine "where is something wrong." In this instance, the scan ruled out pelvic injury and confirmed unusual activity in the ribs, which was consistent with rib trauma (which may or may not have produced fracture). The radiologist included his impression that the rib trauma was compatible with the history of a recent MVA. The impression goes on to mention increased activity in the lateral aspect of the right tibial plateau (the region of the skin directly below the knee). This could have been produced by a knee impact with the instrument panel and therefore would be "compatible" with the patient's history of a motor vehicle accident. It is noted that a bone scan also may provide insight regarding whether an injury (e.g., a fracture) is recent (i.e., metabolically active) or is old.

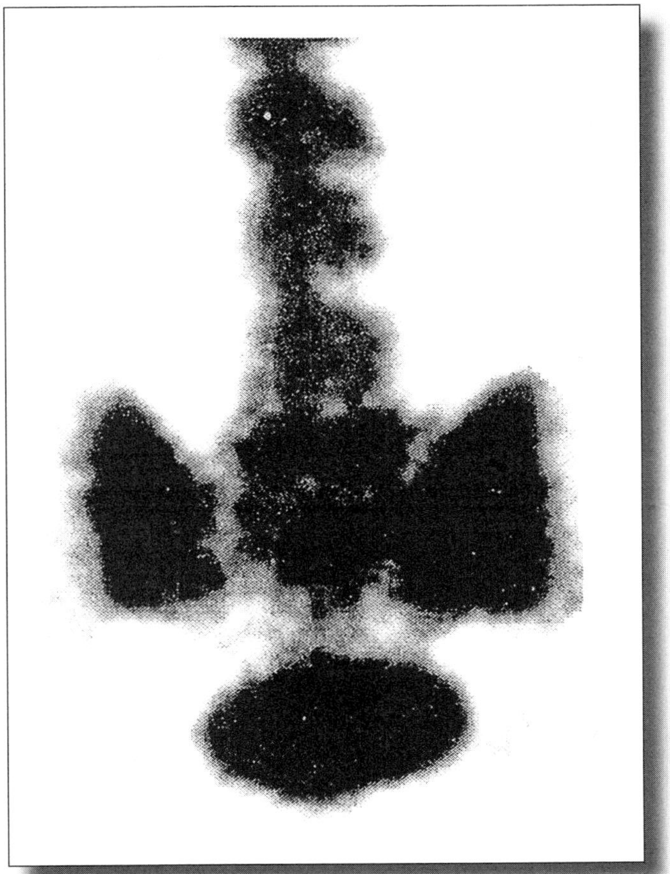

Figure 2-17 *Bone scan—close-up.*

IMPRESSION:

MULTIPLE FOCAL AREAS OF INCREASED ACTIVITY IN THE ANTERIOR RIBS BILATERALLY AS DESCRIBED. THESE MULTIPLE AREAS OF INCREASED ACTIVITY LIKELY REPRESENT SUBACUTE RIB TRAUMA. THIS IS COMPATIBLE WITH THE PATIENT'S HISTORY OF RECENT MOTOR VEHICLE ACCIDENT. THERE IS ALSO INCREASED ACTIVITY IN THE LATERAL ASPECT OF THE RIGHT TIBIAL PLATEAU.

Figure 2-18 *Bone scan radiology report.*

This next report (Figures 2-19 and 2-20) indicates how ultrasound may be used to image blood vessels. In this instance, the patient was immobilized as the result of lower extremity injury, and there was concern about blood clot formation. The report concludes that there was no evidence of deep vein thrombosis (DVT) in either lower extremity. Note that the Conclusion section states that the right side lower thigh could not be imaged because of the presence of a cast. However, because the immobilized side is more likely to have a DVT, that is the side that may have been more productive to image.

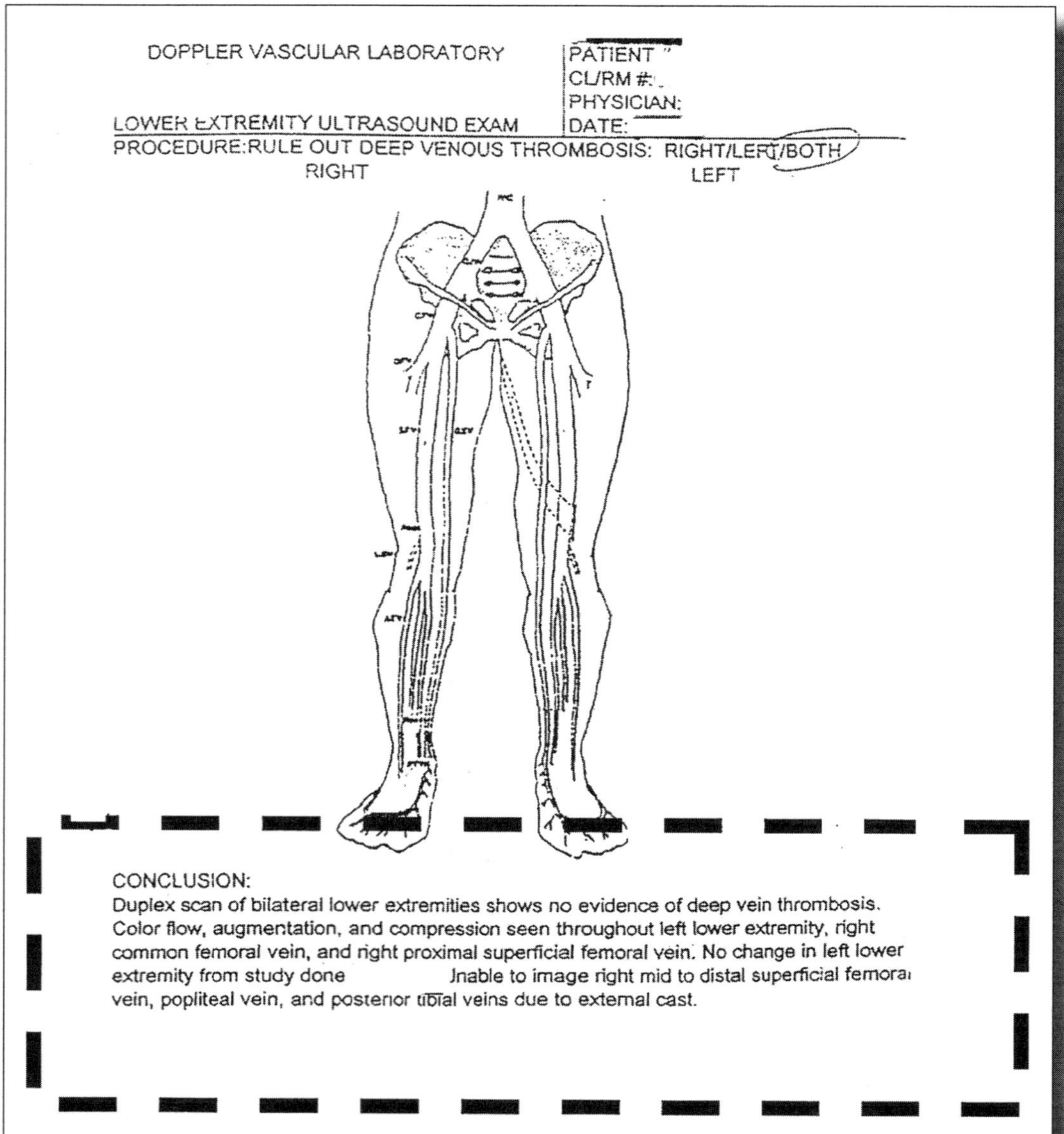
DOPPLER VASCULAR LABORATORY
PATIENT
CL/RM #:
PHYSICIAN:
LOWER EXTREMITY ULTRASOUND EXAM
DATE:
PROCEDURE:RULE OUT DEEP VENOUS THROMBOSIS: RIGHT/LEFT/BOTH
RIGHT LEFT

CONCLUSION:
Duplex scan of bilateral lower extremities shows no evidence of deep vein thrombosis. Color flow, augmentation, and compression seen throughout left lower extremity, right common femoral vein, and right proximal superficial femoral vein. No change in left lower extremity from study done Jnable to image right mid to distal superficial femoraı vein, popliteal vein, and posterior tibial veins due to external cast.

Figure 2-19 *Ultrasound of the lower extremity.*

CONCLUSION:
Duplex scan of bilateral lower extremities shows no evidence of deep vein thrombosis. Color flow, augmentation, and compression seen throughout left lower extremity, right common femoral vein, and right proximal superficial femoral vein. No change in left lower extremity from study done _______ Unable to image right mid to distal superficial femoral vein, popliteal vein, and posterior tibial veins due to external cast.

Figure 2-20 *Ultrasound of the lower extremity—close-up.*

The next document (Figure 2-21) provides another example of a study using ultrasound, in this instance to study the heart. This echocardiogram was performed after trauma produced a cut of an upper chamber of the heart. The cardiogram uses color to display direction and magnitude of blood flow. In this instance, the report indicates that function appears normal. However, the physician indicates that there is no previous echocardiogram available for comparison and, if clinically warranted, a "closer look" could be obtained using a transesophageal echocardiogram (i.e., placing the transducer closer to the heart via the esophagus).

SUMMARY:

The M-mode, two-dimensional and Doppler echocardiographic studies appear normal. Left ventricular function is normal. The left atrium appears normal without significant regurgitant lesion. If more detailed clinical information is warranted post-atrial laceration, a transesophageal echocardiogram could be obtained.

COMMENT: There is no prior echocardiogram currently available for comparison.

Figure 2-21 *Ultrasound echocardiogram.*

This next document (Figure 2-22) is a radiology report of a shoulder MRI, with finding of rotator cuff tear. The rotator cuff is a group of four muscles and the associated tendons that help to support and move the shoulder. The report notes a Type II (inwardly curving) acromion and degenerative changes at the acromion–clavicle junction. The Type II acromion is associated with less space for tendon movement during shoulder abduction (i.e., raising the arm to the side), and degenerative changes (e.g., bony spurs) also can reduce the available space. This, in turn, can make a person prone to rotator cuff injury over a period of time (Jobe, 1998; Steinbach, 1998).

MRI

Patient:
Ref Physician: DOB:
Exam: MRI RIGHT SHOULDER
Exam Date:

CLINICAL HISTORY: Motor vehicle accident four days ago. Rule out rotator cuff tear.

IMPRESSION: FULL THICKNESS ROTATOR CUFF TEAR.

THE STATUS OF THE INTRA-ARTICULAR PORTION OF THE BICEPS TENDON IS UNCERTAIN.

FINDINGS:
Oblique sagittal images demonstrate a type II acromion. There is anterior subacromial spurring and degenerative change at the AC joint.

T2 weighted images demonstrate a full thickness supraspinatus tendon tear with some degree of tendon retraction. Tear diameter is 2 cm. No muscular atrophy of the supraspinatus is noted.

The tendon of the long head of the biceps is identified in the bicipital groove but the status of the intra-articular portion of the tendon is uncertain. The possibility of a partial tendon tear from the labral anchor cannot be ruled out.

No capsular abnormalities are demonstrated.

Figure 2-22 *Right shoulder MRI.*

Note that although a radiology report may distinguish degenerative or pre-existing conditions from recent trauma injury (e.g., Figure 2-9, Impression), the radiologist generally has no information regarding the patient's occupation, sports participation, or other factors. This lack may be significant, to the extent that it affects perceived causation and hence treatment. Also note that the clinician who is providing treatment often relies on the history obtained from the patient. The patient may not recall a particular event accurately, may recall it differently at different times, or may not consider it worth mentioning. Furthermore, the history provided to the clinician may have no information regarding the nature of the crash (e.g., severity, direction, fall). Thus, based on a partial and/or inaccurate history, a clinician may draw reasonable but incorrect conclusions regarding injury causation. Pre-existing conditions are discussed in more detail in Chapter 5.

A number of pre-existing conditions (e.g., previous injury) may make a significant contribution to a patient's chief complaint but may be eclipsed in a history that minimizes or omits them and emphasizes more recent events. The previous contributors can include previous trauma, degenerative conditions, or other diseases. Previous contributors also can include vocations or avocations that require repetitive movements (e.g., overhead lifting that involves frequent movement of structures such as tendons, within a confined space, either congenitally narrow or the site of degenerative changes that reduce available space). The resulting friction produces irritation, inflammation, and perhaps flaying. These cumulative microtraumas are not limited to a particular body region and may include the jaw (temporomandibular joint, or TMJ), the shoulder (rotator cuff, shoulder impingement), superior labrum anterior-posterior (SLAP) lesions, scalene triangle (thoracic outlet), knee (meniscus), and wrist (carpal tunnel). In some instances, a single previous microtrauma or seemingly unrelated event (e.g., heart attack) may initiate a chronic pain syndrome (Zohn, 1998). When these conditions are first noted or treated after trauma, there can be question regarding the extent (if at all) to which the trauma caused the condition or exacerbated a pre-existing condition. Thus, a biomechanical perspective may provide additional insight into the cause and mechanism of injury.

Chapter 3

א

Tests

Introduction

This chapter will focus on three kinds of tests: (1) blood tests, (2) neurophysiologic exams, and (3) psychological exams. In the context of this discussion, our study of blood tests will address the presence of alcohol and/or illegal drugs. Our study of neurophysiologic exams, such as nerve conduction velocities (NCVs), will focus on whether or not there is a change in the electrical functioning of the nervous system that would provide a basis for deficits in mobility or sensation. Finally, our study of neuropsychological testing will focus on whether or not there is a residual deficit in mental functioning as a result of a traumatic event.

Blood Tests

One aspect of blood testing, whether for clinical or forensic purposes, is to determine the blood alcohol concentration (BAC). Primarily, the alcohol concentration will be expressed as a gram percent. Using this convention, the legal limit in all states currently is 0.08. As recently as 2004, some states had a higher limit of 0.10, but the lower limit of 0.08 has now been adopted by all 50 states. Blood also may be tested for other judgment- or behavior-modifying substances such as cocaine. Because mere possession of these substances is illegal, forensic test results usually are reported as positive (present) or not, without specifying the level.

Neurophysiologic Exams

Neurophysiologic exams such as NCVs provide a measure of a nerve's ability to conduct electrical signals. Generally, injury to the nerve or to the myelin sheath surrounding the nerve will present as a slower conducting velocity. This slower conduction may be determined by comparison to population norms and/or by comparison of left versus right for an individual.

Psychological Exams

Neuropsychological testing may be used to detect and specify cognitive deficits. Typically, "before" testing is not available for comparison with "after" testing; therefore, "after" test results are compared to population norms and are interpreted in the context of other "before" information, such as educational level and occupation. Because half the population is, in fact, "below average," a low value per se is not diagnostic of a post-traumatic deficit. Thus, if a 30-year old whose formal education stopped with the third grade was tested and had scores that were somewhat below average, this would not be surprising and would not necessarily indicate that there had been some type of traumatic deficit. In contrast to this, if someone whose educational background included graduate-level college studies scored below average, the contrast between before and after levels of performance would provide a much stronger basis for considering the existence of a post-traumatic deficit (absent other pathology).

Discussion

Blood test results typically will be provided in one of two ways. The first way is as a notation within written medical records such as a progress note, and the second way is as a formal blood test report. An example of each of these follows.

Note that the circled area of this physician's progress record (Figure 3-1) contains the notation "EtOH = 122." The physician is reporting that the ethyl alcohol (EtOH) level was 122 (mg/dL), or, expressed in units consistent with a legal limit of 0.08, would be equal to 0.122. Thus, this patient's alcohol level was about one and one-half times the legal limit.

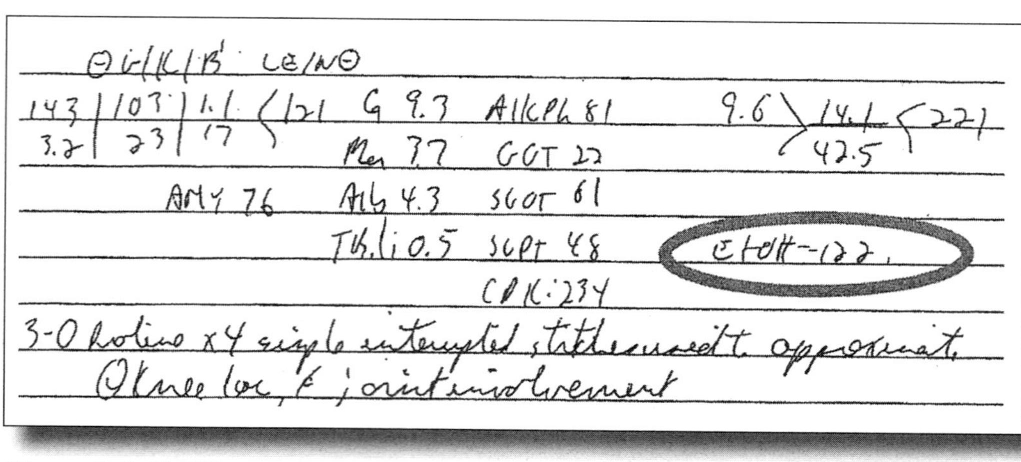

Figure 3-1 *EtOH on a physician's progress report.*

The official alcohol report (Figure 3-2) test results for a different patient indicate an alcohol level of 0.24—three times the legal limit. Note that this is an official report from a crime laboratory. Therefore, in this instance, the sample was taken and assessed for evidentiary rather than treatment purposes.

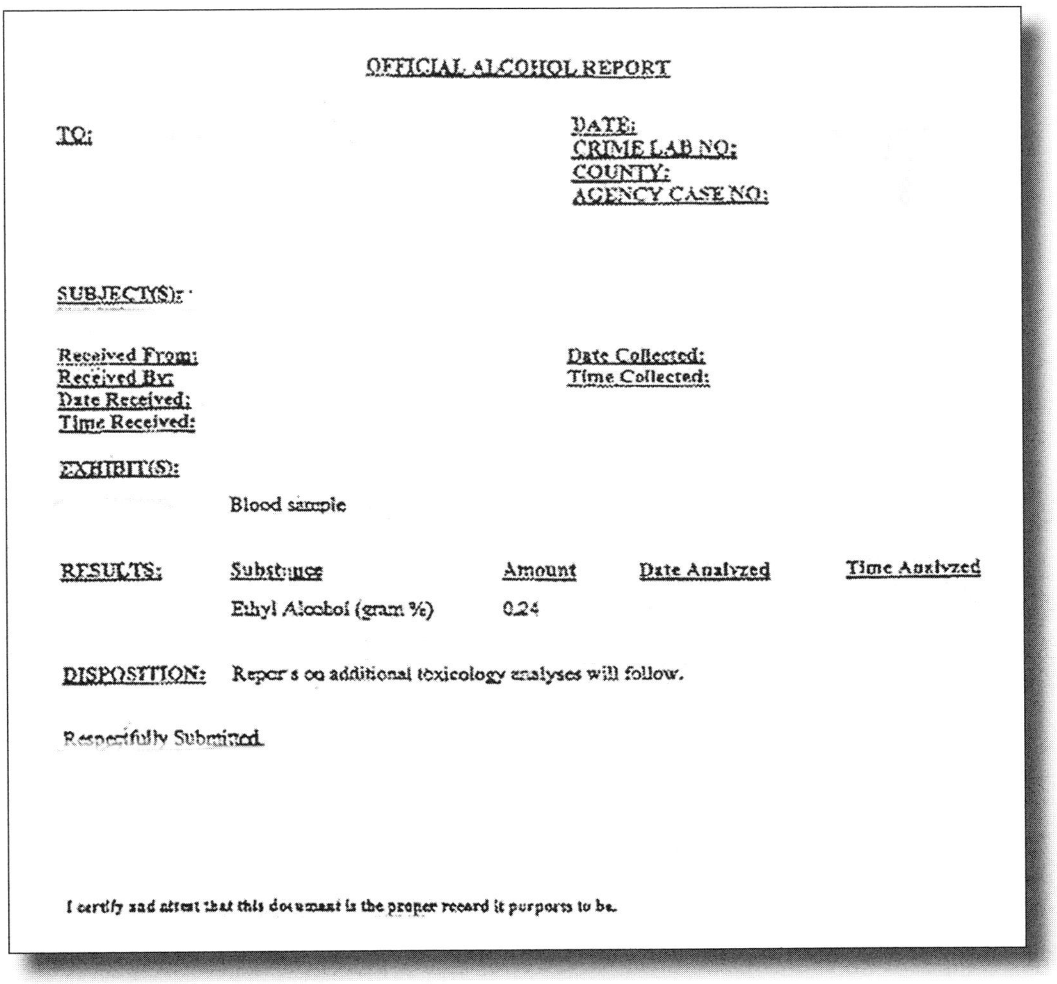

OFFICIAL ALCOHOL REPORT

TO:

DATE:
CRIME LAB NO:
COUNTY:
AGENCY CASE NO:

SUBJECT(S):

Received From:
Received By:
Date Received:
Time Received:

Date Collected:
Time Collected:

EXHIBIT(S):

Blood sample

RESULTS:	Substance	Amount	Date Analyzed	Time Analyzed
	Ethyl Alcohol (gram %)	0.24		

DISPOSITION: Reports on additional toxicology analyses will follow.

Respectfully Submitted

I certify and attest that this document is the proper record it purports to be.

Figure 3-2 *Official alcohol report.*

The next document, an illegal drug report (Figure 3-3), presents results from both blood and urine samples. The first group of test results (blood sample) indicates the presence of a cocaine metabolite, and the second group of test results (urine sample) confirms the presence of cocaine. Note that this sample was taken and assessed for evidentiary rather than treatment purposes.

02500187-1	Blood sample	
02500187-2	Urine sample	
RESULTS:	Substance	Amount
02500187-1	Ecgonine methyl ester	Positive
	9-Carboxy-11-nor-delta-9 THC not detected	
	Negative barbiturate EMIT screen	
	Negative benzodiazepine EMIT screen	
	Negative opiate EMIT screen	
02500187-2	Ecgonine methyl ester	Positive
	Cocaine	Positive
	Cocaethylene	Positive
	9-Carboxy-11-nor-delta-9 THC not detected	
	Negative barbiturate EMIT screen	
	Negative benzodiazepine EMIT screen	
	Negative opiate EMIT screen	

Figure 3-3 *Illegal drug report.*

This NCV study (Figures 3-4 and 3-5) assesses nerve function in terms of the velocity at which it conducts electrical impulses. Because the conduction is being measured over a given length, rather than comparing the velocities per se, one may compare the conduction times. In this instance, rather than comparing this patient to population norms, the clinician was able to make a left-right comparison for the given patient. Note that the top two lines (Figure 3-5), giving values for the left and right median nerves, have approximately the same time (24.38 versus 23.83 ms). However, when examining the bottom two lines, note that the time interval for the left ulnar nerve is considerably less than the time interval for the right ulnar nerve (25.16 versus 27.34 ms). Thus, the right ulnar nerve is conducting rather slowly. This is reflected in the physician's Impression section (Figure 3-6), namely, "Prolonged Right Ulnar F-wave latency is evident." This is associated with nerve root dysfunction on the right side, at the C6/C7 level.

Chief Complaint: Neck pain radiating to the right upper extremity with numbness and paraesthesia

Nerve Conduction Report:

Motor Nerves

Nerve	Site	Onset Lat (ms)	Amplitude	Area (mVms)	Duration (ms)	Seg Name	Delta (ms)	Distance (cm)	Velocity (m/s)
L Median	APB		O-P (mV)	Neg	Neg		O		
	Wrist	3.47	9.62	38.630	6.52	Elbow-Wrist	4.22	26.00	61.6
	Elbow	7.69	9.52	38.455	6.61				
R Median	APB		O-P (mV)	Neg	Neg		O		
	Wrist	3.52	10.70	33.085	5.30	Elbow-Wrist	4.27	26.00	61.0
	Elbow	7.78	10.32	32.034	5.30				
L Ulnar	ADM		O-P (mV)	Neg	Neg		O		
	Wrist	3.05	10.31	34.320	6.66	Elbow-Wrist	4.50	26.00	57.8
	Elbow	7.55	9.15	31.275	6.94				
R Ulnar	ADM		O-P (mV)	Neg	Neg		O		
	Wrist	3.66	9.01	30.630	5.53	Elbow-Wrist	4.36	26.00	59.6
	Elbow	8.02	9.23	29.360	5.77				

Sensory Nerves

Nerve	Site	Onset Lat (ms)	Peak Lat (ms)	Amplitude	Area (µVms)	Duration (ms)	Seg Name	Delta (ms)	Distance (cm)	Velocity (m/s)
L Median 2	2ndDig			P-T (µV)	Neg	Neg		P		
	Wrist	3.06	3.84	40.69	0.033	1.75	Wrist-2ndDig	3.84	16.00	41.6
R Median 2	2ndDig			P-T (µV)	Neg	Neg		P		
	Wrist	3.31	3.94	52.93	0.012	1.06	Wrist-2ndDig	3.94	16.00	40.6
L Ulnar	5thDig			P-T (µV)	Neg	Neg		P		
	Wrist	2.91	3.69	32.17	0.015	1.38	Wrist-5thDig	3.69	16.00	43.4
R Ulnar	5thDig			P-T (µV)	Neg	Neg		P		
	Wrist	2.81	3.66	47.39	0.022	1.41	Wrist-5thDig	3.66	16.00	43.8

F/H Report:

Nerve	Muscle	Lat1 (ms)	Lat2 (ms)	Lat2 - Lat1 (ms)	Amplitude (µV)
L Median F	APB	3.13	27.50	24.38	
R Median F	APB	3.91	27.73	23.83	
L Ulnar F	ADM	3.83	28.98	25.16	
R Ulnar F	ADM	3.36	30.70	27.34	

EMG Report:

Side	Muscle	Nerve	Root	INS	FIBS	PSW	AMP	DUR	POLY	REC	IP	COMMENT
L	CervParaC5-7			Nml	0	0	Nml	Nml	Nml	Normal	Nml	
R	CervParaC5-7			Inc	1+	2+	Nml	Nml	Nml	Normal	Nml	
L/R	Biceps	Musc	C5-6	Nml	0	0	Nml	Nml	Nml	Normal	Nml	

***Figure 3-4** Nerve conduction study (NCS).*

F/H Report:

Nerve	Muscle	Lat1 (ms)	Lat2 (ms)	Lat2 - Lat1 (ms)	Amplitude (μV)
L Median F	APB	3.13	27.50	24.38	
R Median F	APB	3.91	27.73	23.83	
L Ulnar F	ADH	3.83	28.98	25.16	
R Ulnar F	ADH	3.36	30.70	27.34	

Figure 3-5 *Nerve conduction study (NCS)—close-up.*

Impression:
The above electrodiagnostic findings are consistent with right nerve root dysfunction at the level of C6-C7. Prolonged Right Ulnar F-wave latency is evident. Clinical follow up is suggested.

Figure 3-6 *Nerve conduction study (NCS) Impression section.*

Neuropsychological Evaluation

This evaluation (Figure 3-7) is for a man who has a number of low scores on a battery of neuropsychological tests. As is often the case with neuropsychological evaluations, a formal "before" assessment was not available. Thus, the "after" assessment was compared to population averages and other indicators of pre-incident function, such as education and occupation. In this instance, some test results were low, based on both population norms and what would be expected based on the patient's education and occupation. However, the psychologist who is interpreting the results notes that the pattern of the low test results is not typical for post-concussive type of deficits. Indeed, she states that the pattern of deficits is basically the opposite from what one would expect if the deficits were the aftermath of a concussive head injury. The clinician then raises the possibility that the deficit pattern may be indicative of early dementia.

This neuropsychological evaluation report, prepared by another psychologist (Figure 3-8), indicates that the evaluation has taken into account the patient's history as well as the test results, and that overall, it is consistent with diffuse axonal injury (i.e., severe head injury, generally thought to be related to head rotation and typically associated with coma). The evaluation identifies an injury mechanism, namely, coup-contrecoup, that is, brain injury

From a diagnostic standpoint, this pattern is not consistent with expectations for a post-concussive syndrome. In contrast, his performs best on tasks that are often associated with post-concussive syndrome. He shows excellent concentration, abstract reasoning, and attention to detail.

The pattern of deficits suggests that it would be wise to monitor this patient closely to rule out any possibility of an early dementia. While his symptoms are quite subtle and qualitative in nature, they are surprising given the patient's ______ background and current occupation.

Figure 3-7 *Neuropsychological evaluation I.*

CONCLUSIONS

Presence of Injury Due to Brain Dysfunction: Based upon the results of this neuropsychological evaluation and related history, Mr. ______ does show neuropsychological test scores indicative and consistent with diffuse axonal injury secondary to coup-contra-coup phenomena. There are indications of bilateral frontal lobe injury, as well as indications of "spotty injury throughout". Recent MRI findings do indicate "several small foci of injury" which is consistent with diffuse axonal injury.

Figure 3-8 *Neuropsychological evaluation II (citing radiology report).*

associated with head motion that produces brain injury at the site of the head impact (coup) and/or brain injury opposite the site of the head impact (contrecoup). The conclusions quote recent magnetic resonance imaging (MRI) findings of "several small foci of injury." The neuropsychological report does not cite a particular reference for these quotes; however, for instructional purposes, the quotations will be compared to the wording in the MRI report (Figure 3-9). Note that the MRI report describes the foci as "tiny," not small. The two documents (neuropsychological evaluation and radiology report) have another fundamental difference. Whereas the psychologist opines that the foci are "consistent with diffuse axonal injury," the radiologist states the foci are "of indeterminate etiology" (i.e., the foci cannot be ascribed to diffuse axonal injury or any other specific etiology). In fact, the radiologist opines that it otherwise was "an essentially negative cerebral MRI."

OPINION: 1. Several tiny foci of increased signal in the white matter of both the centrum semiovale and corona radiata bilaterally as discussed above of indeterminate etiology. There are perhaps 5 of these tiny foci.

Figure 3-9 *Radiology report (MRI) cited in evaluation report.*

Chapter 4

א

Field Reports

Introduction

Medical records, especially from emergency medical services (EMS) and the emergency room (ER), may contain information regarding seat belt usage and airbag deployment. In some cases, this information may be based on direct observation by EMS personnel at the scene (e.g., the driver was belted, and the belt had to be cut off the driver to extract him). In other instances, the records may be based on "hearsay" (e.g., the occupant told the EMS that she or he was belted, and the EMS repeated it to the ER nurse). Thus, statements in the medical records can have a wide range of bases and validity. Sometimes, the records contain a phrase such as "Pt states that" to indicate the record is a transcript of what the patient said, rather than a conclusion or observation on the part of the care provider.

Narratives often will contain a statement such as "restrained occupant," which refers to the seat belt at that particular seating position being in use at the time of the motor vehicle accident (MVA). Alternatively, the narrative may contain the phrase "positive [+ with a circle around it] seatbelt," "positive restraint," or "positive airbag." As the number of inflatable devices in a vehicle increases (e.g., side airbags and rollover curtains), these notations are likely to become more specific (e.g., positive side airbag) and complex, and perhaps less uniform.

The medical records also may include statements that address belt usage indirectly by providing details regarding possible interactions between the seat belt and the occupant and/or the vehicle interior. For example, if a significant crash is primarily to the front of the vehicle, it would not be surprising to find that the patient had belt-related bruises where the belt's restraining force was applied to the patient. Typically, this would be found along the diagonal from the shoulder to the contralateral hip (e.g., left shoulder to right hip) and across the pelvis, ideally low on the hips, that is, between the anterior superior and anterior inferior iliac crests. (When seated, this typically would be 2 to 3 inches below the navel.) The medical records may include rather detailed description, such as "extensive transverse abdominal bruising [for a standing patient, this bruising would be horizontal across the abdomen], about 2 inches in diameter, located about 3 inches inferior to umbilicus [3 inches below the navel]." Or they may combine few details with an opinion, such as "belt bruising across left abdomen." It is important to note that abdominal bruising is not always indicative of belt usage (e.g., abdominal bruising could be due to abdominal contact with the lower steering wheel rim). Conversely, belt usage does not always produce belt bruising (e.g., the person may have been wearing heavy winter clothing.

As another example of indirect information relating to belt usage, the medical records may indicate that there was a spider-web-like fracture pattern on the windshield, little vehicle damage (usually a marker of a relatively low-speed impact), and forehead bruising or laceration. It is likely that the head–windshield contact produced the head bruising and windshield fracture, and this would be unlikely to occur if the person were belted. Similarly, if the person sustained knee bruising in this impact, it most likely would be caused by knee impact with the lower instrument panel. This also would be unlikely if the person were belted. Of course, this also would depend on other information, such as the height of the person and whether the seat were full forward or back. The farther back the seat was positioned, the less likely a belted occupant would be to strike his or her knees. This additional information may not be provided in the medical records. However, there may be a comment from those at the scene regarding deformation of the lower instrument panel, which, in conjunction with a low-speed impact, an average or tall patient (who is more likely to have had the seat adjusted full rearward), and knee impact injury, points toward a scenario in which the belt was not worn. Similarly, the medical records may contain information that indirectly indicates airbag deployment. For example, a record of some reddening in the lower face, neck, upper chest, and/or inside of the forearms may correlate with airbag deployment. The medical records also may provide some insight into restraint usage/deployment by the injuries that are not recorded. Seat belts help to reduce injury severity in a wide range of crashes—including frontal, side, rear, and rollover—and airbags help reduce head and neck injury in moderate and severe frontal crashes. This section includes reports from various personnel in the field or the transition from field to facility (e.g., EMS personnel, police, hospital admissions personnel, ER personnel). There also are a few examples of how these "primary sources" are reflected and utilized by clinicians "further down the chain." The records may be prepared by professionals who are considered impartial (e.g., EMS), or they may be prepared by professionals who are considered as advocates (e.g., a caregiver typically may be a patient advocate, but in a particular instance may be defensive about a treatment that did not have an optimum result).

Discussion

The first record is the narrative section of an ER record (Figure 4-1) and may be read as follows:

> *MVA [motor vehicle accident], C/C [chief complaint] is neck pain. [Age is blocked out, patient is about 50.] __ year old female, c/o [complains of] neck pain after rear end MVA yesterday. Positive for seatbelt [that is, she was wearing her seat belt at the time of the accident].*

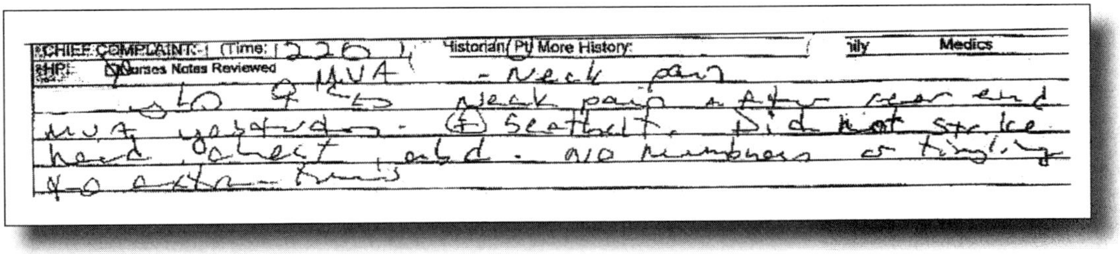

Figure 4-1 *Positive for seat belt.*

Note that at the top of the page, it specifies "Historian" as Pt (patient), indicating the historical information was obtained from the patient. The entry regarding seat belt usage appears to be part of the information the patient is providing, rather than some type of independent observation by the care provider. The record continues:

> *Did not strike head, chest, or abdomen. No numbness or tingling to extremities.*

The narrative states that the MVA occurred the preceding day and makes no mention of the patient seeking or receiving medical attention in the past day. Therefore, it seems reasonable to assume that this is her first contact with the healthcare system since the MVA occurred. Because there is no mention of EMS and so forth, a reasonable assumption is that the history was provided by the patient directly to the care provider. Indeed, this agrees with the notation at the top of the narrative, indicating that the patient was the historian. Thus, the history should be read with the caveat of any notes that are not extemporaneous and may have some embedded subjectivity. The narrative does not directly address severity, but if it was a very low-speed rear impact, the question of seat belt usage may not be significant.

The next record is a consultation report (Figure 4-2), that is, a specialist was asked to contribute to the care of this patient. His notes are very clear with regard to belt usage and airbag deployment. The record states "restrained passenger" and "positive for airbag deployment." What is less clear is the basis for writing "restrained," that is, whether he was repeating a history provided by the patient or an impression conveyed by EMS personnel. If this was an impression from EMS personnel, was it based on EMS direct observation, or was it based on something the patient or a bystander told them? Similarly, the statement regarding "airbag deployment" is of unknown source or basis. For example, this patient may have been a passenger occupying a seating location that did not have an airbag, even though an airbag did deploy in the vehicle.

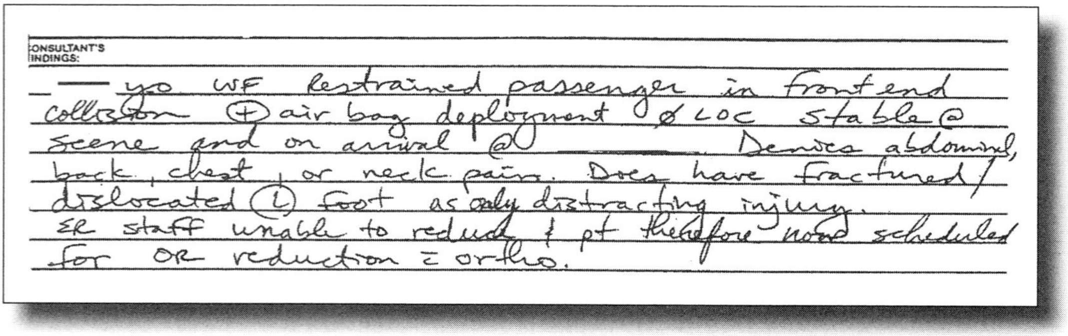

ONSULTANT'S
INDINGS:

— yo WF restrained passenger in front end
collision ⊕ air bag deployment ∅ LOC stable @
scene and on arrival @ ___ Denies abdominal,
back, chest, or neck pain. Does have fractured/
dislocated Ⓛ foot as only distracting injury.
ER staff unable to reduce & pt therefore now scheduled
for OR reduction c̄ ortho.

Figure 4-2 *Airbag deployment.*

As an exercise, the Findings section may be read as follows:

> *"__ year old white female, restrained passenger in front end collision, positive for airbag deployment, no loss of consciousness (LOC), stable at [the] scene and on arrival at __. Denies abdominal, back, chest, or neck pain. Does have fractured/dislocated left foot as only distracting injury. ER staff unable to reduce [i.e., unable to properly align bone fragments], and patient therefore now scheduled for operating room reduction with orthopaedics [i.e., the bone fragments will be surgically realigned in the operating room rather than in the emergency room].*

The next record (Figures 4-3 and 4-4) is from trauma admission. It has the following entry, "ABDOMEN: Epigastric seat belt sign." That is, in the epigastric region of the abdomen (the region surrounding the navel), the care provider observed and noted bruising, which, presumably, because of the shape and location of the bruise, she or he interpreted as being due to a seat belt. Note that the entries for the other body regions tend to be more objective, containing observations or measurements. An entry for the abdominal region that would be more consistent with the objective nature of the other entries might be "epigastric contusion, 1″ wide, 3″ long, transverse."

As an exercise, the rest of the sheet may be read as follows: The heading at the top is Physical Exam and includes an initial measurement of the vital signs (e.g., blood pressure 118/72, and heart rate 72), and beneath that, the Glasgow Coma Scale (GCS) rating of 15. This rating corresponds to normal brain functioning. Lower values correspond to increasingly severe deterioration of mental functioning. (An overview of anatomical terminology and injury scales may be found in Pike, 1990.)

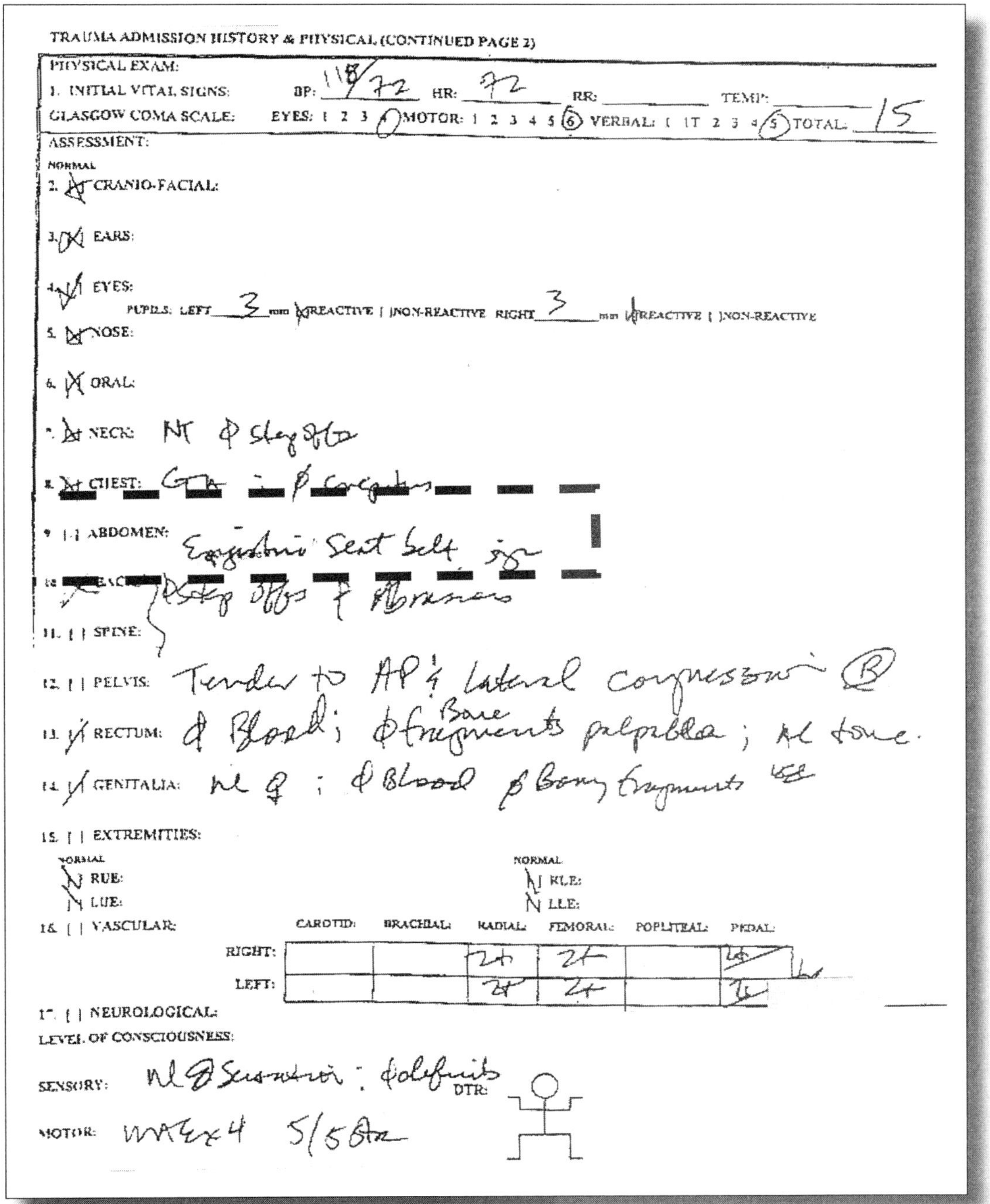

TRAUMA ADMISSION HISTORY & PHYSICAL (CONTINUED PAGE 2)

PHYSICAL EXAM:

1. INITIAL VITAL SIGNS: BP: 118/72 HR: 72 RR: ______ TEMP: ______

GLASGOW COMA SCALE: EYES: 1 2 3 (4) MOTOR: 1 2 3 4 5 (6) VERBAL: 1 1T 2 3 4 (5) TOTAL: 15

ASSESSMENT:

NORMAL

2. [X] CRANIO-FACIAL:

3. [X] EARS:

4. [X] EYES:
PUPILS: LEFT 3 mm [X] REACTIVE [ ] NON-REACTIVE RIGHT 3 mm [X] REACTIVE [ ] NON-REACTIVE

5. [X] NOSE:

6. [X] ORAL:

7. [X] NECK: NT Ø step offs

8. [X] CHEST: CTA [redacted] Ø crepitus [redacted]

9. [ ] ABDOMEN: Epigastric seat belt sign [redacted]

10. [redacted] BAC[redacted] Ø step offs Ø abrasions

11. [ ] SPINE:

12. [ ] PELVIS: Tender to AP & lateral compression (B)

13. [X] RECTUM: Ø Blood; Ø Bone fragments palpable; nl tone.

14. [X] GENITALIA: nl ♀ ; Ø Blood Ø Bony fragments

15. [ ] EXTREMITIES:

NORMAL
[N] RUE:
[N] LUE:

NORMAL
[N] RLE:
[N] LLE:

16. [ ] VASCULAR:

	CAROTID:	BRACHIAL:	RADIAL:	FEMORAL:	POPLITEAL:	PEDAL:
RIGHT:			2+	2+		1+
LEFT:			2+	2+		1+

17. [ ] NEUROLOGICAL:

LEVEL OF CONSCIOUSNESS:

SENSORY: nl Ø sensation: Ø deficits

DTR:

MOTOR: MAE x 4 5/5 str

Figure 4-3 Seat belt sign.

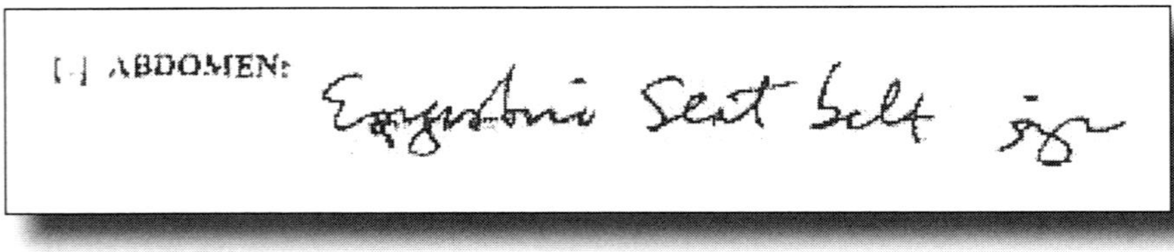

Figure 4-4 *Seat belt sign—close-up.*

Moving down to the Assessment section of Figure 4-3, we see numerous other remarks:

- Eyes: Both pupils react to light (i.e., the pupillary reflex is intact), and the pupils are 3 mm in diameter, meaning they are neither dilated nor constricted.
- Neck: NT (non-tender) with no step-offs (i.e., no displacements that would be indicative of bony injury).
- Chest: No crepitus (a crackling sound accompanying breathing, which, when present, may indicate lung disease or pneumothorax).
- Back: No step-offs, no abrasions.
- Pelvis: Tender to AP (anterior-posterior), and lateral compression on the right side.
- Rectum: No blood, no bone fragments palpable (i.e., no bone fragments detectable by touch), normal tone.
- Genitalia: Normal female, no blood, no bony fragments.
- Extremities: Left upper extremity (LUE), right upper extremity (RUE), left lower extremity (LLE), and right lower extremity (RLE) all normal.
- Vascular: Pulses present and of strength 2+ at the wrist (radial), thigh (femoral), and ankle (pedal), thus indicating there are no fractures or other vascular insults that would interfere with pulses at these key monitoring locations.
- Sensory: Normal with no deficits.
- Motor function: The patient can move all four extremities (MAEx4), and strength is 5 out of 5.

The following provide examples of photographs taken in the field and the types of information they may provide (Figures 4-5 through 4-20). If in-vehicle emergency cameras move beyond the concept stage, such photographic records may become more widespread, more immediate, and more routinely part of the medical records.

This section of the lower instrument panel on the driver's side shows two prominent contact marks (also referred to as "scuff marks") (Figure 4-5). Most likely, these were produced by knee contact, and this should be reflected in the medical records, perhaps by knee bruising or even knee fracture. Note that the axial loading produced by knee impact also may result in femur and/or pelvic fracture. Depending on the specifics of the crash (e.g., direction and severity), knee impact also may be relevant in addressing seat belt usage.

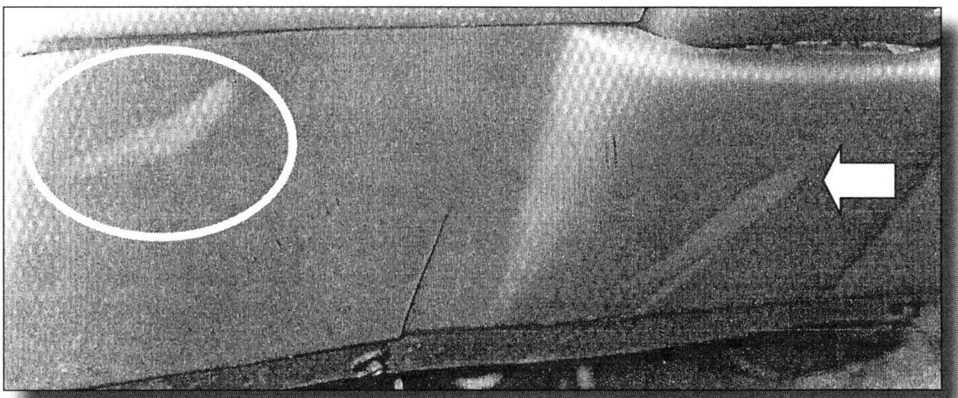

Figure 4-5 *Lower instrument panel, driver's side.*

This next photograph (Figure 4-6) shows a contact mark underneath the glove box handle. As in the previous example, this most likely was due to contact with the occupant's knee and may be reflected as knee bruising, or other knee contact injury, in the medical records.

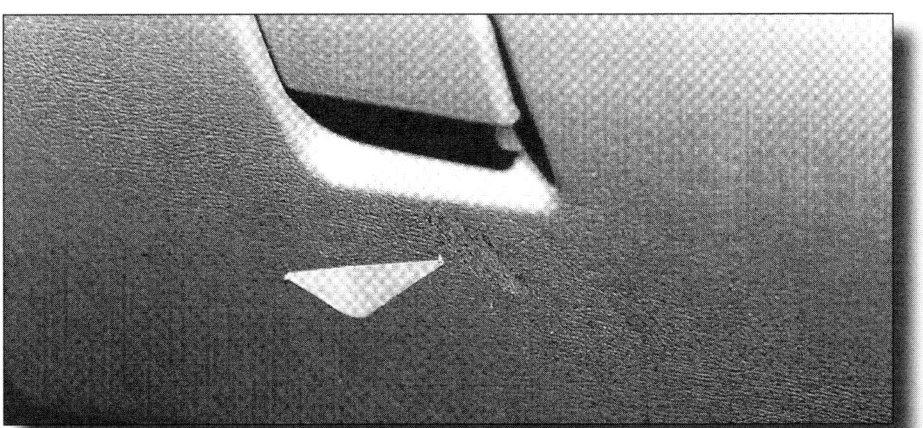

Figure 4-6 *Lower instrument panel, passenger's side.*

The next two figures (Figures 4-7 and 4-8) depict a section of the upper instrument panel near the glove box. The second photograph (Figure 4-8) shows a close-up of the region that was circled in Figure 4-7. The contact marks on the upper instrument panel correlated well with medical records, indicating injured teeth.

Figure 4-7 *Upper instrument panel, glove box region.*

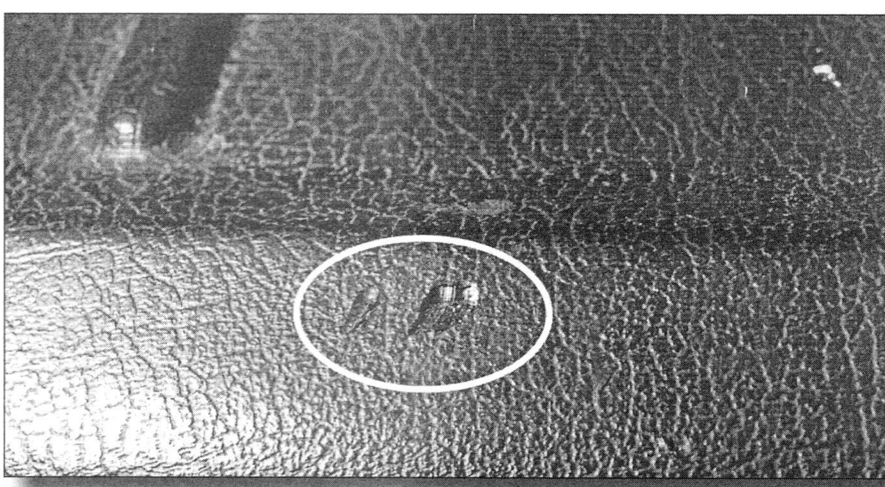

Figure 4-8 *Upper instrument panel, glove box region—close-up.*

The next photograph (Figure 4-9) shows an abraded region of the sun visor. Closer examination, provided in Figure 4-10, shows several hairs in the abraded area. This correlates well with a scalp laceration reported in the medical records.

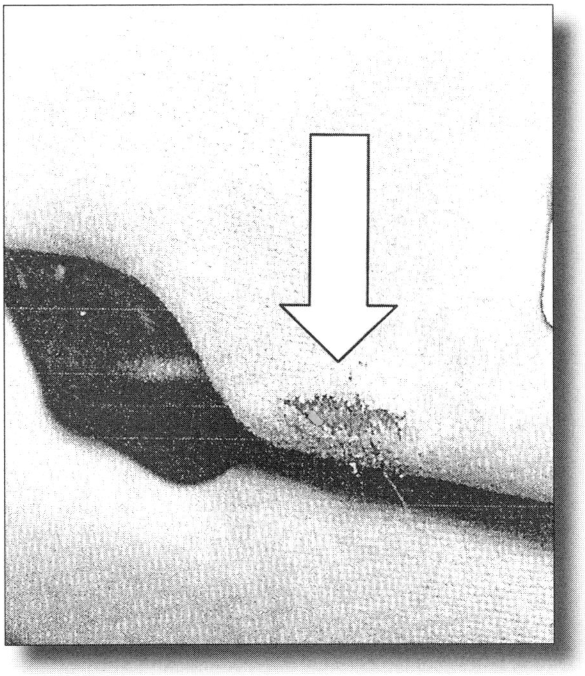

Figure 4-9 *Sun visor.*

Figure 4-10 *Sun visor—close-up.*

The next photograph (Figure 4-11) shows a so-called spider-web fracture pattern that is typical of high-speed impacts involving the windshield and an object such as a head or fist. Note that the fracture pattern is characterized by rather straight, radial lines and a series of concentric circles. The 1-1/2 mark on the "ruler" points to the center of the web, which should coincide with the point of contact. Note that if this impact were ascribed to a head impact, the injury analysis should address the feasibility of a head-shaped object being able to impact that location on the windshield. Also, the impact point may provide valuable insight regarding occupant position and hence seat belt usage.

Figure 4-11 *Cracked windshield.*

The next photograph (Figure 4-12), a close-up of a cracked windshield, shows hair embedded within some of the cracks This correlates well with scalp laceration reported in the medical records.

The next figure (Figure 4-13) shows a side view of a steering wheel. The measuring tape that is resting against the wheel rim also serves as a straight edge and shows that the rim is not deformed (i.e., is not bent forward or backward). Such bending may occur as the result of a number of interactions, including the occupant impacting part of the wheel or the occupant "bracing himself" by locking his elbows and pushing down on the steering wheel rim.

Figure 4-12 *Cracked windshield embedded with hair.*

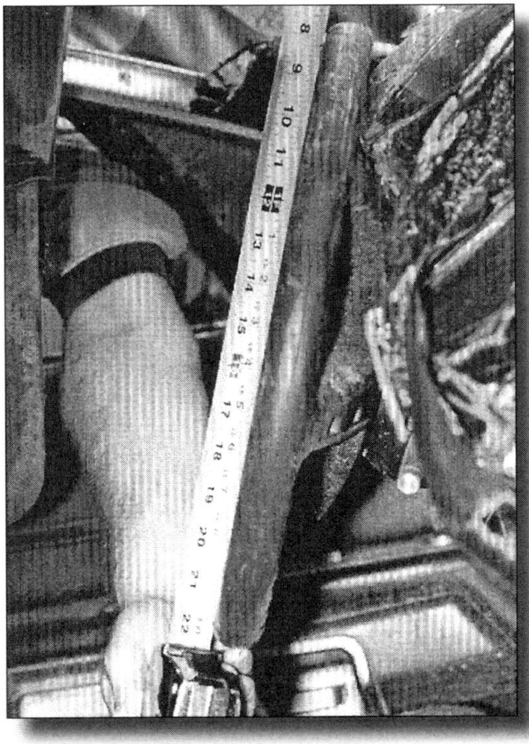

Figure 4-13 *Steering wheel.*

This next photograph (Figure 4-14) shows the metal latch plate or "tongue" of a seat belt. The perspective of the photograph is looking from the webbing toward the tip of the tongue. The arrow points toward one of the surfaces where the buckle typically latches and hence where load marks would be expected if the seat belt were worn during a substantial vehicle impact. In the example shown, there are no load marks.

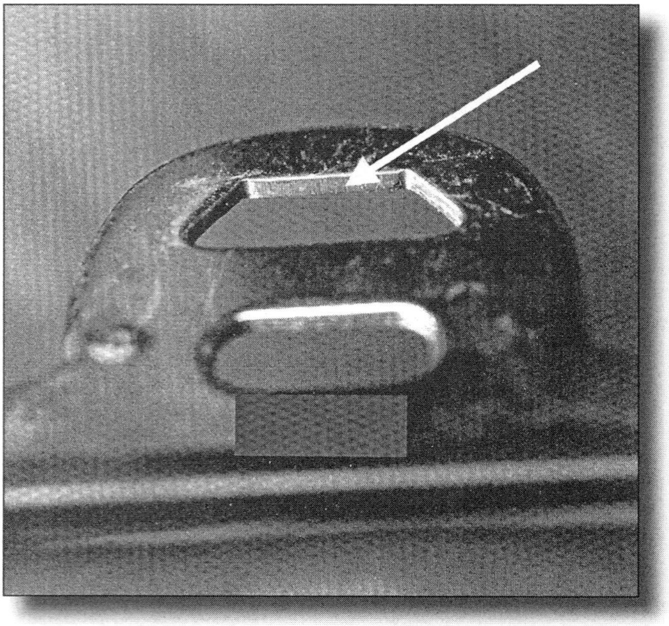

Figure 4-14 *Seat belt latch plate.*

This next photograph (Figure 4-15) shows a seat belt latch plate (tongue), with a vertically oriented usage mark.

This next photograph (Figure 4-16) shows abrasions on the plastic part of the latch plate through which the belt webbing is routed. Typically, these types of abrasions are made by the belt webbing, as loading by the occupant during a crash event stretches the webbing and pulls it through the latch plate. Note that the abrasions are oriented at an angle. This is typical of the orientation of the webbing with respect to the latch plate when the belt is being loaded during a crash.

This next photograph (Figure 4-17) shows abrasions on the plastic D-ring (shoulder belt upper routing point). Typically, these abrasions are caused by the seat belt webbing as loading by the occupant stretches the webbing and pulls it through the D-ring. Thus, this would be an indication that the belt was worn during a crash.

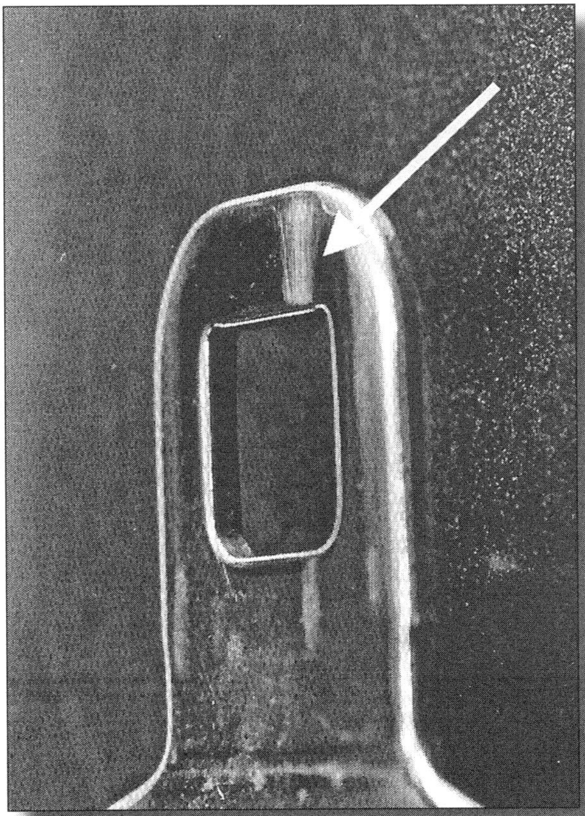

Figure 4-15 *Seat belt latch plate usage mark.*

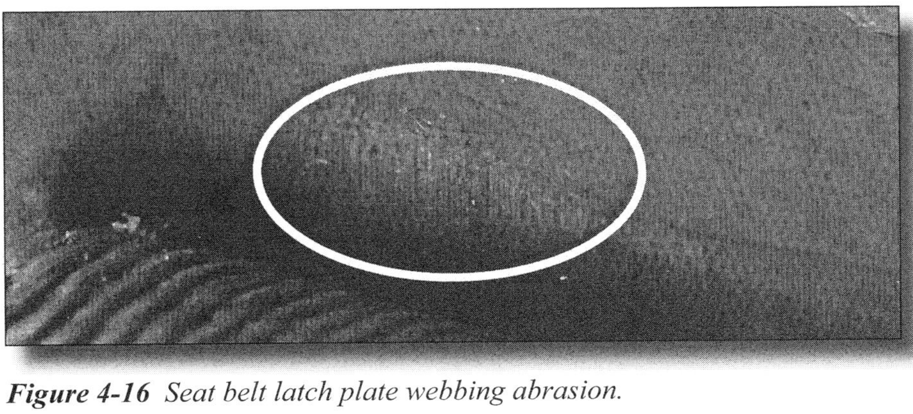

Figure 4-16 *Seat belt latch plate webbing abrasion.*

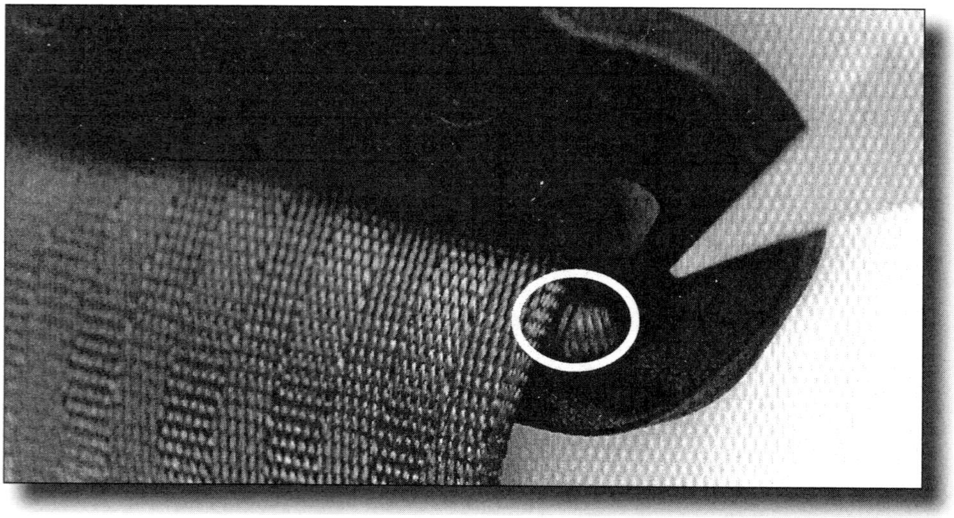

Figure 4-17 D-ring abrasion.

The next photograph (Figure 4-18) shows a load mark on seat belt webbing. Note that the mark is oriented diagonally with respect to the width of the belt, which reflects the angle between the webbing and the latch plate when the belt is loaded during a crash.

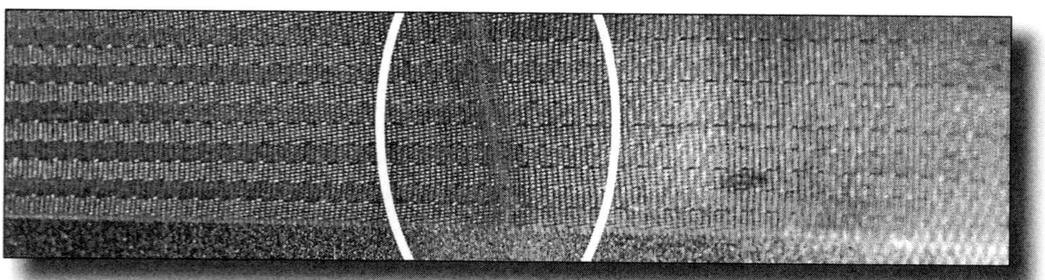

Figure 4-18 Seat belt webbing.

This next photograph (Figure 4-19) shows shoulder and chest bruising. Note that the bruise starts at the right shoulder, is oriented diagonally, and goes downward and to the left. This is consistent with shoulder belt placement for a right front passenger and provides an indication that the person was seat belted during the crash event. Note that the width of the bruise is rather constant and consistent with the width of a seat belt.

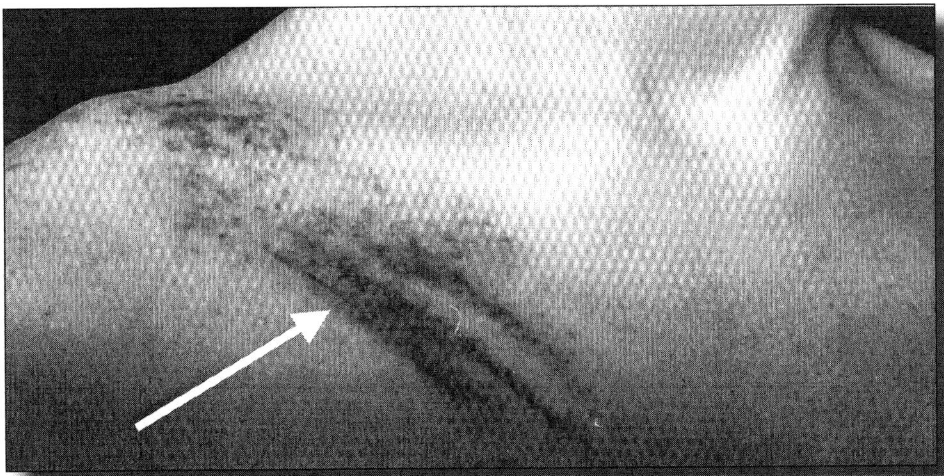

Figure 4-19 *Shoulder bruising.*

This next photograph (Figure 4-20) shows a deployed airbag. The bag is from the right front passenger location and is relatively large (about 24 inches in diameter) compared to the driver's airbag. Also, note what is not shown in the picture, namely, there is no noticeable tissue residue or staining. Especially if someone is wearing makeup, the deployed airbag can be quite stained; in some instances, lip prints may even be discernable on the bag. In the preceding example, there is no indication on the bag of occupant contact. If the medical records indicate no facial, neck, or upper chest contusion, then the vehicle inspection together with the medical records are rather consistent in not supporting significant airbag contact. Thus, other injury mechanisms (i.e., non-airbag) may be more appropriate for any injuries to these regions.

The following section contains a variety of records that either directly or indirectly address seat belt use. Often, a medical record will describe a patient as a belted occupant of a vehicle that was in an MVA; however, the record frequently does not contain much information describing the basis for concluding that someone was belted. For example, was it part of the history provided by the patient, was it per EMS personnel who cut the belt off the occupant as part of the extraction procedure, or was it part of the information that was derived from earlier medical records?

This emergency department report (Figure 4-21) states that the patient was a "back seat passenger" with "no restraints." The report does not specify who provided the information (e.g., was it provided via EMS or directly from the patient), nor does it provide detail regarding whether "no restraints" means that the seat belts had been removed from the rear seat prior to this MVA, if the belts were available but not worn, and so forth.

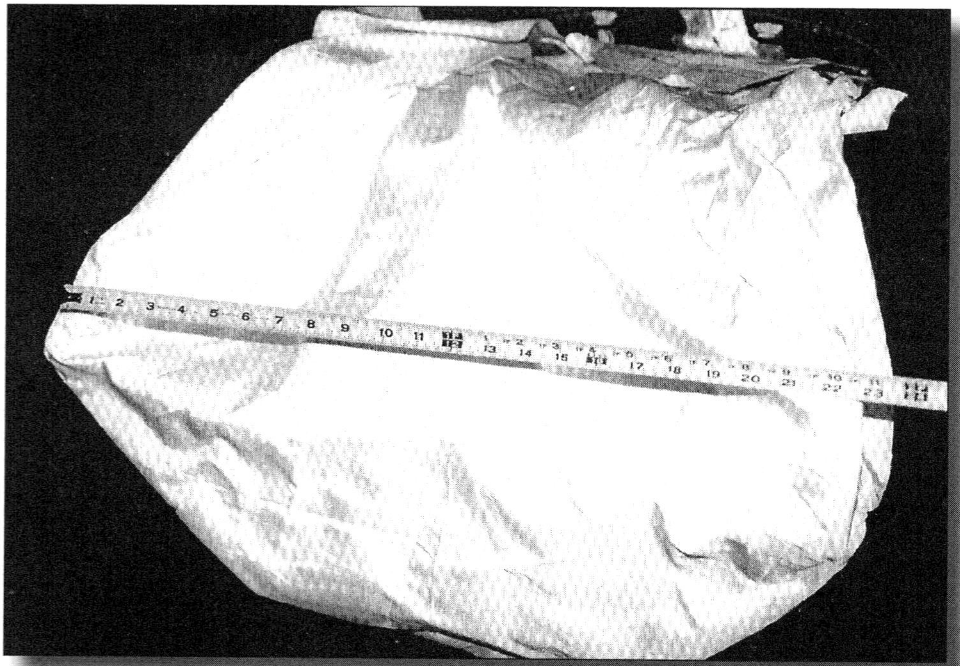

Figure 4-20 *Deployed airbag.*

MVA, BACK SEAT PASSENGER, NO RESTRAINTS. IMMOBILIZED BY [redacted] EMS. TIME

Figure 4-21 *No restraints.*

This next record (Figure 4-22) states that it was "unknown" if the patient was restrained. We do not know from this record if the patient said he or she did not remember, if the EMS conveyed uncertainty, or if the author of the report has doubt, one way or another, regarding seat belt usage. It also is not explicitly stated whether or not an airbag deployed.

This next record (Figure 4-23) states that the patient was "reportedly restrained." The record does not indicate the source who reported this (i.e., the EMS, the patient, the triage nurse). It also is not known if this wording is merely a statement of fact or if the record writer is injecting some skepticism regarding belt use in this instance.

of vehicle unknown if pt was restrained.

Figure 4-22 *Unknown restraint use.*

HISTORY OF PRESENT ILLNESS: The patient is a ——— male who was a passenger in a motor vehicle accident, who was reportedly restrained.

Figure 4-23 *Patient reportedly restrained.*

This hospital record (Figure 4-24), as part of the history, states that the patient was "restrained" (last word of the first line). The history states, "Blunt abdominal trauma with lower abdominal wall tenderness." It is not clear if the abdominal tenderness was the basis for concluding that the patient was belted (i.e., that the tenderness was from restraint force applied by the lap belt against the abdomen). The abdominal tenderness per se may not be a good basis to conclude that the seat belt was used (i.e., the lower abdominal tenderness may have been due to impact loading by the lower part of the steering wheel rim).

DATE:
CC: — y/o WM driver MVA Restrained
HISTORY: Blunt Abd Trauma c̄ Lower Abd
wall Tenderness. No [illegible]
Hematuria, No LOC, Ambulatory at scene

Figure 4-24 *Patient restrained.*

Although this discharge summary (Figure 4-25) does not directly address seat belt usage, the injury that was treated can provide some useful insight. More specifically, there was a "posterior rim acetabular fracture." For a frontal impact, the knees may impact the instrument panel, and the resulting force can push the femur backward. If there is sufficient force, it may "pop" the

> To more definitely confirm that there was no fracture in the joint, a CT scan of the right hip was done postoperatively. This showed that there was a posterior rim acetabular fracture, but it was not in the joint.

Figure 4-25 *Belt use and hip fracture.*

head of the femur out from the hip socket. As this occurs, the head of the femur may press on and fracture the rim of the socket portion of the ball and socket hip joint. Depending on the severity of the crash, other injuries, and load marks, this amount of knee loading more likely may be associated with an unbelted occupant than with a belted one.

This consultant report (Figure 4-26) states that the patient was the "restrained" driver. From the letter, the source of this information is unclear; presumably, it was from other medical records and/or history provided by the patient. The report also includes the statement that "Police would not come to the scene." No further information is provided on this, so one is left to speculate whether the police considered the crash to be minor or if there was another reason.

> Thank you for referring this ________ female who was injured in a motor vehicle accident on ________________. At the time, she was the driver of her vehicle and restrained. She was driving ________ She went through an intersection. The opposing vehicle ran a red light and hitting the right rear of her vehicle, causing it to go into a 90-degree spin. Police would not come to the scene. She exchanged information with the other driver and took a cab home.

Figure 4-26 *Consultant report regarding restraint use.*

This EMS run sheet (Figure 4-27) states that the patient was restrained. It also contains additional detail regarding the crash scenario, namely, that it was a frontal collision with major front-end damage and airbag deployment and that there was "obvious contact" with the upper corner of the windshield. This report is impressive in the useful detail that was recorded. However, note that from a biomechanical analysis perspective, a number of subjective statements were presented as objective. Depending on where in the upper corner of the windshield the impact fractures were located, it may or may not have been feasible for the impact to have involved the head. Furthermore, if the head was able to reach the impact point, it may not be

CASE NARRATIVE F, W RESTRAINED FRONT-SEAT PASSENGER OF FRONTAL COLLISION MVA. MAJOR FRONT-END DAMAGE
c̄ AIRBAG DEPLOYMENT. OBVIOUS CONTACT • UPPER CORNER OF WINDSHIELD. STRIKE IN PASSENGER SEAT.
SEMI-CONS. PALE DIAPHORETIC. (+) MOVEMENT ALL EXTREMITIES. OBVIOUS FACIAL TRAUMA. MID-FACE

Figure 4-27 *EMS notes regarding restrained passenger.*

compatible with the person being belted, that is, depending on the details of occupant size, crash severity, orientation, and seating position, it may not have been feasible for a belted occupant to have struck that region of the windshield with his or her head. The report does not describe glass fracture per se, but it is reasonable to assume that the obvious contact reference involves some windshield fracture. There is no mention of "starring" (i.e., the glass fracture pattern that looks somewhat like a star or spider web). If, for example, the seat was fully rearward and the occupant was very short and sustained two fractured knee caps, there would be a question whether a head–windshield impact could have reasonably been attributed to a belted occupant.

This emergency department (ED) report (Figure 4-28) is somewhat unusual, in that it states, "he states he was restrained, but this is in question." Thus, the ER physician is explicit about the patient being the historian but says that he doubts the accuracy of what the patient has said about wearing a seat belt at the time of the crash event. This ED report then states that the paramedics reported starring of the windshield. Presumably, one of the care providers interpreted the windshield starring as indicative of head contact with the windshield, and this in turn was interpreted as indicating that the occupant was not belted in this crash.

He was the right front passenger (he states he was restrained, but this is in question) of a ____________ vehicle that allegedly swerved to avoid another vehicle and struck a tree. It is not clear whether there was loss of consciousness. According to the paramedics, he was alert on the scene, but the window was starred.

Figure 4-28 *Patient restraint questionable.*

This patient care report (Figure 4-29) contains a great deal of background information regarding the injury. More specifically, the report states that the patient was a restrained front-seat passenger of a vehicle that had a low-speed head-on impact with a tree and that the impact produced 4 to 6 inches of front-end damage. It also provides information regarding the type

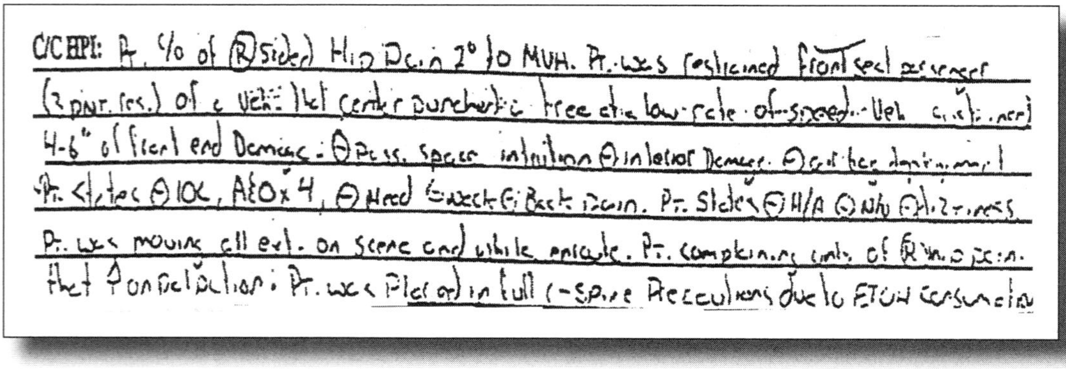

Figure 4-29 *Patient care report regarding restrained patient.*

of restraint used, namely, that it was a "2 pt. res." This term has been used to refer either to a "lap belt only" seat belt configuration, such as that found in the center front seating position of many vehicles, or to the shoulder belt portion of a lap and shoulder belt with a passive shoulder belt (i.e., the lap belt is buckled manually, and the shoulder portion of the belt is automatic). Another record (not shown) indicates that the patient was the right front passenger; thus, in this instance, it appears that the term is being used to describe a passive shoulder belt/manual lap belt configuration and that the lap belt was not used.

Also note that the report states that the patient was administered full cervical protection due to "EtOH consumption," that is, because the patient had consumed alcohol. Presumably, the patient was sufficiently intoxicated to cause the EMS personnel some concern about the patient causing additional injury (e.g., by voluntary movement because the patient was "feeling no pain" and thus might move in a way that ordinarily would be avoided because it would be too painful or perhaps by falling). It also may be appropriate to take the intoxication into account when considering the patient's reliability as a historian and seat belt use.

The next record, part of a police accident report (Figure 4-30), describes a one-vehicle accident involving an impact with a telephone pole. The police officer determined that alcohol was not involved and addresses another possibility, namely, a distracted driver. The first line of the narrative states, "No cell phone present." As the number of potential distractions increases and their possible effect on driving performance is realized, it is likely that newer versions of traffic accident forms will have one or more check boxes to address this.

Unit #1 No Cell Phone Present

This accident occurred as Unit #1 was traveling North ________ Operator #1 lost control of the vehicle and struck a Telephone Pole off the east side of ______ Impact with the pole was made with the right side of the vehicle. Unit #1 then traveled approx. 82'8" before exiting the roadway. Unit #1 traveled off the paved portion of ______ northbound approx. 32'8" before impacting a tree. This was the position of Unit #1 upon this officer's arrival.

Figure 4-30 *No cell phone present.*

Chapter 5

א

Pre-Existing Conditions

Introduction

Typically, a biomechanical injury analysis will seek to identify the mechanism by which an injury occurred and whether or not the injury was caused by a particular event. If too many possible scenarios are included in the analysis, they can obscure the actual injury causation. Therefore, whenever possible, the analysis should be simplified and better focused by eliminating as many alternatives as possible. One basis for elimination is if an injury or condition was pre-existing, that is, in existence before the crash event occurred. In this context, a pre-existing condition can be defined broadly as any process or condition that had its origin before the event being studied. This would include a previous injury, a previous disease process or condition, and a previous condition that required medication that produced relevant side effects. These three categories can be further subdivided to include congenital, degenerative, disease, iatrogenic, injury, intoxication, medication, and nosocomial factors. Note that some or all of these factors may not have been apparent to the individual or to his physicians but may be discernable using a posteriori biomechanical injury causation analysis. This chapter includes examples of various pre-existing conditions and examines some of the different ways they may be reflected in the medical records.

Discussion

Case 1. Pre-Existing Disease

This discussion will focus on the neck region of the spine and will address a pre-existing condition, namely, degenerative change. The radiologist identifies (Figure 5-1) degenerative disc and joint disease (which would not be due to recent trauma) in the middle and lower cervical spine and a slight backward displacement (posterior subluxation) of C5 with respect to the vertebra immediately below it, C6. This subluxation might be interpreted as a recent

CERVICAL SPINE 3 VIEWS:

History: Motor vehicle accident.

FINDINGS: Multi level degenerative disc and joint disease is seen in the mid and lower cervical spine. There is slight posterior subluxation of C5 on C6. This is believed to be degenerative in nature. There is no evidence of acute fracture or dislocation. Prevertebral soft tissues are normal.

Figure 5-1 *Degenerative cervical spine.*

extension-induced injury, but the radiologist provides the additional finding of "This is believed to be degenerative in nature," thus not caused by recent trauma. The patient's age is blocked out, but she is in her 60s; therefore, age-related degenerative changes would not be totally unexpected. The radiologist then notes that "There is no evidence of acute fracture or dislocation," that is, no recently acquired bony injuries.

Case 2. Pre-Existing Injury

This typed patient history (Figure 5-2) is part of a clinical consultation report. The patient was in a motor vehicle accident (MVA) and perceived some deficits developed a while afterward. Although the deficits were detected after the MVA and therefore could be assumed to be caused by it, his medical history indicates another possible basis for the deficits, namely, he previously had a number of major head injuries, including a head injury as a child, a previous MVA, and a number of concussions when he was a boxer.

PAST HISTORY: He's had a number of major head injuries, at ___ y/o he hit his head and lost consciousness, he was a boxer when he was in ________ apparently he had a number of concussions there, he also had a car wreck when he was in and of course, had another car wreck this time all had resulted in his having loss of consciousness. He is not noted to have any medical problems such as diabetes, heart disease or high blood pressure.

Figure 5-2 *Previous head injury.*

CLINICAL HISTORY: INJURY

THREE VIEWS OF LUMBAR SPINE, FRONTAL VIEW OF THE PELVIS AND LATERAL VIEW OF THE RIGHT HIP, THREE VIEWS OF THE RIGHT FEMUR

FINDINGS: There is generalized osteopenia. There is narrowing of the disc space between L5 and S1. IVC filter is seen in place. No evidence of fracture or dislocation.

The pelvis and right hip demonstrates significant degenerative changes and osteopenia with status post hardware fixating pelvic fracture and femur fracture. No evidence of acute fracture.

The patient is status post intramedullary rod fixating a previous femur fracture. No evidence of acute fracture.

Figure 5-3 *Previous femur fracture.*

This radiology report (Figure 5-3) addresses previous injury and degenerative condition. These are x-rays of the lower back, pelvis, and right thigh. The radiologist notes osteopenia, a general term for decreased bone density. Possible causes for the decrease include osteoporosis, osteomalacia, and other metabolic bone disorders (Boden, 1991). Note that the patient's age (although the exact age is blocked out in Figure 5-3, the patient was under 30 years of age) points away from age-related osteoporosis. The radiologist also notes hardware in place to hold together a previous pelvis and femur fracture but adds that there is no evidence of "acute fracture," that is, no new fractures (due to the most recent trauma). This radiology report provides no additional information regarding the bone loss per se; however, given the earlier trauma, the bone loss is likely due to pre-existing disease, previously administered medication, and/or previous trauma. This can be further analyzed by noting that the previous fracture would likely limit weight-bearing and hence mobility on the (previously) injured side and thereby foster decreased bone density due to decreased usage. Thus, the osteopenia most likely was due to the mobility reduction while the previous injury was healing. Note that it would take a number of months after mobility was restored before the bone density "recovered." It is noted in passing that the radiologist observed a filter implanted to restrict the passage of possible blood clots, as the blood flows through the inferior vena cava (IVC).

Case 3. Pre-Existing Intoxication

This record (Figure 5-4) is part of the ER or ED (Emergency Room or Emergency Department) record and addresses intoxication and belt usage. The section labeled "Brief History" may be read as follows:

> *C/C [chief complaint] of MVA [motor vehicle accident]—unrestrained driver ejected out. Found in back seat. [Thus, the EMS determined that the person who was found in the back seat when EMS arrived had actually been the driver*

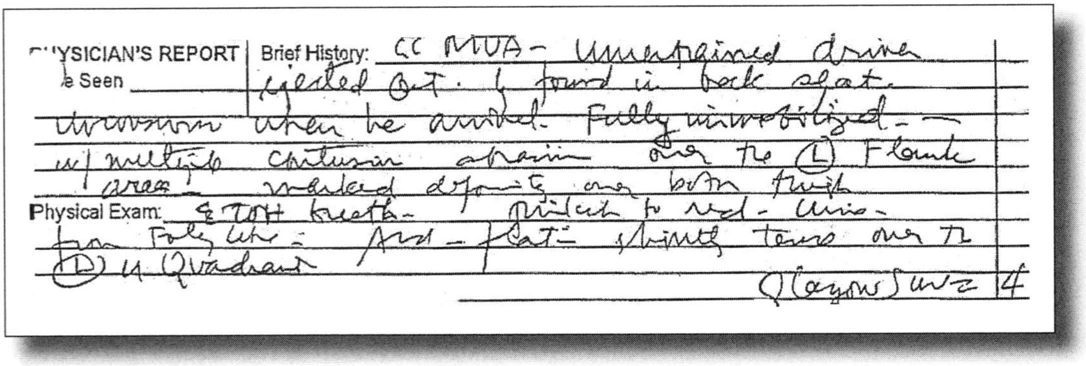

Figure 5-4 *Intoxication.*

> *and was thrown into the back seat area as part of the crash.] Unconscious when he arrived [via ambulance to the emergency room]. Fully immobilized with multiple contusions, abrasions over the left flank area [left side]. Marked deformity over both thighs [both femurs probably fractured]. EtOH breath [ethyl alcohol smelled on breath]. Pinkish to red urine from Foley catheter, abdomen flat, slightly tense over left upper quadrant. [The abdomen can be divided conceptually into four quadrants, with origin at the umbilicus.]*

Note the notation in the lower right-hand corner, "Glasgow Score = 4." The Glasgow Coma Score (GCS) provides a measure of head injury, and a value of GCS = 4 corresponds to a very severe decrease in brain function. (The GCS is discussed in Pike, 1990.)

This next range of motion study (Figure 5-5) includes the results of an MRI exam of the shoulder, which found a Type II acromion. Pre-existing structure may cause or pre-dispose to repetitive use injury (sometimes referred to as "cumulative microtrauma"). In this instance, a part of the shoulder, the acromion, was shaped in such a way as to be confining and to cause excessive frictional wear of the tendons and other soft tissue during everyday shoulder movements. In particular, tears of the supraspinatous tendon (part of the rotator cuff) may be associated with the more restrictive acromion structures (Steinbach, 1998). The clinician describes the joint as having "hypertrophic arthritic changes." Thus, the joint irregularities are not attributable to the MVA. Another possible scenario would be that the MVA exacerbated some of the pre-existing conditions.

This next report (Figure 5-6) of a radiographic exam notes that the exam was degraded by motion artifact, that is, the images were somewhat blurry due to movement during the imaging. Apparently, however, the radiologist had sufficient image quality to form the impression that there were severe abnormalities of the AC (acromioclavicular) joint and that these changes were degenerative in nature (i.e., occurring gradually rather than acute or suddenly). Although this section of the report does not address timing, the imaging was performed relatively soon after the trauma. Thus, there would not have been sufficient time for the trauma

> Three views of lumbar spine reviewed, reveal partial lumbarization of S1 vertebral body. There is diffuse anterior osteophyte throughout lumbar spine, no evidence of fracture dislocation or soft tissue swelling.
>
> MRI of left shoulder reveals an intact rotator cuff, there is Type II acromian and some acromioclavicular joint hypertorphic arthritic changes.

Figure 5-5 *Type II acromion.*

> IMPRESSION: Exam degraded by motion artifact as discussed above.
>
> Full-thickness tear of the rotator cuff involving the supraspinatus and the infraspinatus tendons.
>
> Severe degenerative changes of the acromioclavicular joint.
>
> Slight superior subluxation of the humeral head with narrowing of the acromiohumeral distance. There is also slight widening of the anterior aspect of the glenohumeral joint which may reflect instability or disruption of the anterior capsule. Clinical correlation is suggested.
>
> The biceps tendon is probably intact as discussed above, but evaluation is somewhat limited due to the motion artifact. Clinical correlation is suggested.

Figure 5-6 *Degenerative acromioclavicular joint.*

to have resulted in the degenerative changes. Rather, the degenerative process was already underway at the time of the trauma.

This next patient had pre-existing heart disease (Figure 5-7) and was wearing a cardiac monitor at the time of this MVA. She had a "distracting" cardiac event while driving and, as a result, temporarily lost control of the vehicle and impacted a tree. Thus, in this instance, rather than the crash producing chest compression and heart injury, it was the pre-existing heart disease that caused the vehicle crash.

> You may recall this _ y/o woman was hospitalized a few weeks ago following a motor vehicle accident which may have been preceded by transient loss of consciousness or transient inability to sustain control of her car. I saw her in the hospital due to the concern for possible arrhythmia which may have preceded the accident. She was also in atrial fibrillation, which was newly diagnosed for which the patient had been recently started on _____ prior to the accident.

Figure 5-7 *Previously reduced cognition.*

In this follow-up report (Figure 5-8) to an office visit, the physician is providing his impression regarding a trauma patient with pre-existing disease. Due to the pre-existing disease, there are records of radiological and electrophysiological testing before the crash; due to the trauma, there are results of similar testing after the crash. Although the testing shows changes on the "post-trauma" tests compared to the "pre-trauma" tests, the physician basically is saying that the timing with regard to the trauma is coincidental, rather than there being a causal relationship between the trauma and the changes. He notes that there are changes to some areas of the brain, but that this is due to long-standing circulatory disease involving the small blood vessels of the brain. He also addresses to what the brain lesions are not due, that is, "There is no evidence that the patient sustained head injury as part of the accident per se." He then states that he does not "attribute the changes on the scan or the changes on the EEG [electroencephalogram] to her motor vehicle accident." Thus, if viewed on a time line, the intracranial changes would appear to be due to the MVA; however, in fact, the two merely were correlated and not causally related. Also note that the lack of closed head injury may be relevant regarding whether the associated neck motion was extreme enough to be a likely cause of cervical spine injury.

> IMPRESSION/PLAN
>
> The patient has radiographic and electrophysiological evidence of some ischemic cerebrovascular disease of small vessel nature. She has a long history of high blood pressure for which she sees Dr. [redacted] There is no evidence that the patient sustained head injury as part of the accident per se and I do not attribute the changes on the scan or the changes on the EEG to her motor vehicle accident. The patient continues to have spinal symptoms of cervicothoracic and lumbar sprain/strain syndrome.

Figure 5-8 *Before and after comparison of test results.*

This next case (Figure 5-9) was somewhat unusual, in that the patient's injuries were significantly affected by pre-existing disease of someone not involved in the crash. At the time of the crash, the subject vehicle was outfitted with hand controls for the spouse, and it was hand striking and scraping against the hardware of these hand controls that caused the patient's hand injury. It is interesting to note that later medical records included histories that the patient had injured her hand on the vehicle interior but had no mention that the interior had been significantly altered with the addition of hand controls.

REPORT OF CONSULTATION — y.o w ♀, involved in MVA, ? LOC airbags deflated. Struck hand possibly on hand controls (for husbands)

Figure 5-9 *Indirect effect of pre-existing hardware.*

This hospital discharge summary (Figure 5-10) discusses a complex case, with a variety of previous conditions and events that may have influenced the current condition. More specifically, the patient had been hospitalized (for three months) for treatment of encephalitis, after which he had deficits of both cognition and mobility. (These are both significant with regard to future events in terms of his "very weak" ambulatory skills, making falls more likely and his cognitive deficits perhaps affecting his reliability as a historian.) The patient then was

HISTORY: [redacted] is a [redacted] year old [redacted] male with a history of herpetic encephalitis in The patient was subsequently hospitalized at under the care of Dr. [redacted] His hospitalization was for approximately three months. He subsequently did have problems with problem solving skills as well as ambulation, describing his ambulation initially as being "very weak". The patient [redacted] was involved in a motor vehicle accident with a closed head injury and loss of consciousness for approximately fifteen minutes. He did not require hospitalization at that time. On [redacted] the patient was admitted to [redacted] with a history of that on the morning of [redacted] he struck his head on the refrigerator door with subsequent headache and weakness of the entire side of his body. The patient was found to have suffered a large right hemispheric complex subdural hematoma and on [redacted] Dr. [redacted] performed right frontotemporal parietal craniotomy,

Figure 5-10 *Subsequent injury.*

involved in an MVA that produced brief unconsciousness but did not require hospitalization. Somewhat after the MVA, the patient struck his head on a refrigerator door, which produced headache and unilateral weakness. (The patient was alone before the refrigerator incident; thus, it is not known if he had previous refrigerator impacts and/or falls that in the aggregate contributed to his current condition.) Although a hematoma was detected after the refrigerator incident, it is not known if this collection of blood was forming gradually over time, in which case its origin pre-dated the refrigerator impact.

This next consultation report (Figure 5-11) includes a discussion of various medical records that indicate hyperextension injury of the neck, myofascial syndrome, and a pre-existing disease, arthritis, that was aggravated by a traumatic event.

Dear Dr. ________

I saw ________ in the office today for follow-up visit. Since last seen, she went to the ________ There, she was treated and diagnosed. Their diagnosis was hyperextension injury of the neck, myofascial syndrome and traumatic aggravation of pre-existing arthritis.

Figure 5-11 *Arthritis aggravation.*

This report (Figure 5-12) summarizes a physician's thinking regarding if an injury is new or old. Based on previous records, including a bone scan and spine x-rays, there is a mixture of old and new injuries. The compression of the T12 vertebra was documented on imaging taken before the current incident, but the compression of the T5 vertebra is new and is presumed attributable to the recent trauma.

A new x-ray is obtained today and the compression of T5 does not appear to be significantly altered. In addition, I have had a chance to review some previous records which include old bone scan and other x-rays of the thoracic and lumbar spine. She has a history of a T12 compression which was confirmed by bone scan, however she has no abnormality at T5 indicating that this is a new injury based on the recent accident.

Figure 5-12 *Old versus new imaging.*

This next narrative report (Figure 5-13) discusses a complex history, including previous motor vehicle accidents that required treatment for some of the current complaints—neck and back problems. In addition, the patient's history includes placement of a pacemaker and, at the time of the current MVA, use of prescription medications, including an anti-depressant. Thus, the previous injuries may have contributed to her current complaints, and medication taken to treat pre-existing conditions may have contributed to the occurrence of the current MVA. (From the record provided, we cannot tell if the patient was the driver.)

PAST HISTORY

The patient has been involved in previous accidents, including in ____ as a pedestrian she was struck by a as she was crossing the street, in ____ of ____ she was struck by a ________ truck, and in ____ of ____ she was struck by a ________________. The patient was previously treated and released from care for neck and back problems relating to the ____ and ____ motor vehicle accidents.

The patient's past surgical history includes the placement of a pacemaker on ____________. The patient is taking medication at this time in the form of ____________ and Zoloft.

Figure 5-13 *Final narrative report.*

The next report (Figures 5-14 and 5-15) discusses two vertebral fractures. The C5 fracture (Figure 5-14) was not produced by the recent trauma or by a previous trauma, but rather the bone broke as a result of a recent surgery. The C7 fracture (Figure 5-15) appeared to present on pre-event x-rays, but the surgeon did not feel the "before" x-rays were conclusive. To be on the safe side, he treated the fracture as if it were new.

The surgery went well, and he was discharged home to convalesce. He returned in ______ with a set of routine postoperative films, and they demonstrated that there was a fracture through the body of C5, but it was not displaced. This is a complication, but one that occurs postop.

Figure 5-14 *Old versus new fractures, part I.*

> Emergency Room. A radiologist identified a radiolucent line in the C7 vertebral body, and there was a question raised whether this was an acute fracture. I reviewed the old studies, ones which were not available to the ______ and it is my opinion that this was probably not a recent fracture. We immobilized him and treated him as though it were because we could not be 100% sure, but on the other hand, when I reviewed the old films, it looked as though this radiolucent line was probably already present.

Figure 5-15 *Old versus new fractures, part II.*

This next report (Figure 5-16) describes a complex history including disc herniations at C5/C6 and C6/C7 that were treated surgically, an iatrogenic fracture of the C5 vertebra, and more recently, an MVA, after which additional surgery was required and performed. The writer opines that the additional surgery was attributable to the recent MVA.

> I supplied him with pain medicines for a while, but then informed him that he would have to find another doctor in view of the fact that no surgical considerations were in order. In summary, he did have a disk herniation at the C5-6 and C6-7 levels that was treated surgically. His postoperative course was complicated by a C5 vertebral fracture. This occurred spontaneously. He subsequently was involved in a car accident which materially worsened his condition. That led to the development of recurrent left arm pain for which a repeat operation was necessary, namely a microhemilaminotomy which was performed on ______ It is my opinion that this second operation was a result of the automobile accident. It is also my opinion that, when last seen, he had a dysesthetic-type pain syndrome which was caused by the automobile accident.

Figure 5-16 *Previous disc rupture and surgery.*

This ER report (Figure 5-17) discusses an incident of a patient reportedly being choked at the site of a previous neck surgery. It is not clear from the document provided how much of his current complaint involves the natural course of his previous condition and treatment and how much is attributable to the choking incident.

HISTORY OF PRESENT ILLNESS/REVIEW OF SYSTEMS: This __ year-old man had been drinking alcohol with his friends over the last several hours, and for whatever reason one of his friends grabbed hold of him around his throat, anterior neck area, and squeezed. The patient is complaining of feeling like there is something in his throat, kind of a gaggy feeling, irritation. He was concerned because he has had several surgeries on his cervical spine, has a plate in the cervical spine. He has not coughed or brought up any blood. No real shortness of breath.

Figure 5-17 *Choking incident.*

This emergency department (ED) record (Figure 5-18) describes a fatality that apparently had several contributing factors. The patient's vehicle left the road and was involved in a crash that produced a liver injury, from which he apparently bled to death. However, this occurred against a backdrop of end-stage COPD (chronic obstructive pulmonary disease, a condition that could have been fatal in its own right) and a possible heart attack (which may have precipitated the crash). In addition, the patient was intoxicated; the level is not specified, but presumably the level was enough to impair the patient's ability to avoid and then to recover from the vehicle trauma.

ruptured his liver and that he has bleed out from there. He had what appears to be end-stage COPD, did not tolerate lying on his back or having his large abdomen push on the diaphragm as evidenced by his PC02 of 66. This is a guy who drove off the road with an elevated ST segment on the monitor, he may have gone off the road because of a heart attack. It is also felt possible that the stress of what he was doing through, is a factor to this. The alcohol intoxication also was felt not to help things.

Figure 5-18 *Exsanguination.*

Chapter 6

א

Case Studies

Introduction

Typically, the cases that follow include several records per case and discuss integrating these to arrive at an overall impression. The different records often reflect different perspectives, and when appropriate, these are identified and discussed. In some instances, the additional records merely provide additional confirmatory information; in other cases, the information may seem contradictory. In either event, the discussion aims toward arriving at an overall interpretation/impression.

Case Study 1

The first example is derived from a case also discussed in Chapter 1. However, additional comments now will be offered regarding interpretation.

This record is an emergency medical services (EMS) run sheet or run report (Figure 6-1). In many instances, the EMS are among the first people on the scene. They are trained not only in medical care but, as part of that, in observing scene information that may help in treatment.

The report indicates that the vehicle was in a frontal impact and that it sustained substantial damage. This record reports that the patient was "unrestrained," that is, unbelted. Thus, it would not be surprising if the patient impacted the vehicle interior, the so-called second collision. The patient was semi-conscious when found, so the EMS considered a head injury. Furthermore, because it was a frontal impact, the EMS were looking for a windshield fracture as a sign of head impact with the windshield. Typically, a significant head impact to the windshield will produce a spider web or starburst fracture pattern; however, "no starring of the windshield was found." Thus, we can conclude that if there were a head injury that caused the patient to

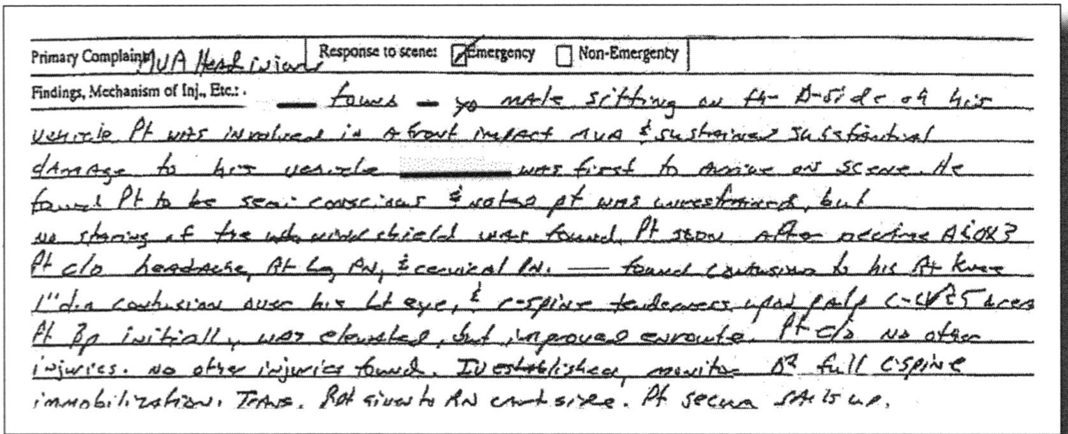
Primary Complaint: MVA Head injury | Response to scene: [x] Emergency [ ] Non-Emergency

Findings, Mechanism of Inj., Etc.: [redacted] found [redacted] yo male sitting on the D-side of his vehicle. Pt was involved in a front impact MVA & sustained substantial damage to his vehicle. [redacted] was first to arrive on scene. He found Pt to be semi-conscious & states pt was unrestrained, but no starring of the windshield was found. Pt soon after declined A&Ox3. Pt c/o headache, Rt leg, Rt, & cervical pn. — found contusion to his Rt knee 1" dia contusion over his Lt eye, & c-spine tenderness upon palp C-4/C5 area. Pt Bp initially was elevated, but improved enroute. Pt c/o no other injuries. No other injuries found. IV established, monitor O2 full C-spine immobilization. Trans. Rpt given to RN cont. size. Pt secure status up.

Figure 6-1 *EMS run sheet, with no windshield starring.*

be semi-conscious at the scene, it probably was due to the crash causing a general jostling of the head rather than from a head impact (at least not an impact with the windshield).

Case Study 2

This case involves a vehicle that slid across a roadway and came to rest near a tree, possibly after impacting the tree. The traffic accident report (TAR) (Figure 6-2) states that the passenger complained of hitting her jaw on the passenger-side window.

The next record, a radiology report, is for a complete series of facial bone x-rays. This report (Figure 6-3) had two findings: a slightly depressed fracture of the left zygomatic arch (cheek bone) and a minimal diastasis (partial separation) of the left frontal-zygomatic articulation. These findings led to an addendum being added to the medical record (Figure 6-4), which might otherwise have been interpreted as indicating that all suspected injuries had been ruled out.

The addendum includes a summary of the radiographic information from a neck x-ray and a CT scan of the head. The cervical spine was read as within normal limits, and the head computed tomography (CT) scan was read as negative. Without knowledge of the facial injuries, one might reason that the lack of neck injury would be one indicator that there was no extreme neck motion produced by this MVA, and the CT negative for intracranial bleeding again would be an indicator of no extreme head acceleration. Furthermore, although the CT was performed specifically to rule out intracranial bleeding, one might expect the report to note the presence of a skull or facial fracture. This presumed lack of fracture also would have supported the notion that there was no extreme head impact.

DRIVER #1 was driving eastbound
on ___ and when reaching
___ lost control of his
vehicle on ice covered road. His
vehicle slid across ___
stopping in the yard by a tree
A ___ ___ passenger
hit her jaw on the passanger
side window ___ ___

Figure 6-2 *Traffic accident report, jaw impact.*

Multiple views reveal a slightly depressed fracture involving the left zygomatic arch. There is also minimal diastasis of the left frontal-zygomatic articulation.

Figure 6-3 *Facial x-rays.*

ADDENDUM

This _-year-old ___ comes to the emergency room with chief complaint of being in an auto accident. Dr. ___ examined the patient initially and ordered X-rays and CAT scan which were negative. She is waiting for the plain films, and came back showing fracture of the zygomatic arch with clouding of the left maxillary sinus.

Figure 6-4 *Medical record addendum.*

This patient was reportedly the right-front passenger in a frontal impact; therefore, one might be surprised that she struck the side window. However, the medical records did not detail "how frontal" the impact was. Thus, there may have been a significant lateral component. Be that as it may, the patient was the right-side passenger, and one might expect injury due to side window contact to involve the right side, rather than the left side, of her head. One important factor in this is the patient's position at the time of the impact. The records presented do not address the patient's orientation pre-impact, but one explanation would be if the patient were turned in her seat, such as might be involved in looking back over her right shoulder, to speak with a rear-seat occupant or to retrieve an object from the back seat.

The next record (Figure 6-5), which is from the report written by a plastic surgeon, provides a good example of a history indicating that it is not necessarily the observations or impressions of the writer but rather clearly stating that this history was told to the writer and who told it. The history, as recorded by the surgeon, relates the following: the passenger-side airbag did not inflate (the history does not explicitly address whether or not the driver-side airbag deployed), and although the patient was wearing the lap and shoulder belt, she hit the windshield.

> The history that I received from ________ was that she was a restrained passenger in an automobile. The car skidded on some ice and the front of the car was hit. Although the vehicle was equipped with air bags, it did not inflate on the passenger side, and although she had a seatbelt with shoulder harness on, she apparently hit the windshield. Because of difficulties in opening her mouth, an oral surgeon was called. Since the injuries were eventually determined to be facial bone fracture, she was referred for plastic surgical management.

Figure 6-5 *Plastic surgeon's report.*

Note a small but significant difference in the history regarding whether the patient impacted the side door glass or the windshield. These would be very different scenarios regarding injury mechanism and occupant kinematics. We do not know if the history as told had changed, or if this reflects a change accidentally introduced along the way.

Case Study 3

This report (Figure 6-6) of an electrocardiogram (EKG) indicates a fast heart rate and conduction abnormality compatible with cardiac ischemia. Thus, the report is saying there is

FINAL INTERPRETATION: Sinus tachycardia: T wave inversions compiatible with but not diagnostic of cardiac ischemia.

Figure 6-6 *EKG.*

abnormality that could be due to the heart not getting adequate blood supply. However, the results are not conclusive, in that the abnormality may have other causes as well: the clinician is saying that she or he has not ruled out cardiac ischemia but cannot yet make a diagnosis of cardiac ischemia.

Case Study 4

This case involves a long and complex history of complaints involving dizziness. As noted in the report from the ear, nose, and throat specialist (ENT) (Figure 6-7), the testing has included an audiogram, ENG (electronystagmogram), and brain magnetic resonance imaging (MRI). The audiogram was normal, but the ENG revealed abnormal eye movements, possibly indicative of middle ear involvement. Furthermore, the MRI revealed at least three possible blood vessel malformations (arterio-venous malformations [AVMs]). The clinician indicates that he is unclear how these findings relate to the complaints of dizziness, which may be related to migraines or the malformed vasculature and has referred the patient to a neurosurgeon for further evaluation.

Since [redacted] last visit, we have obtained an audiogram, ENG, and an MRI of the brain. The audiogram was essentially within normal limits. The ENG, however, did reveal some abnormal saccadic testing, as well as some abnormal pursuit, indicative of possible central vestibular involvement. The MRI scan revealed at least three small possible AVM's as noted on the report.

I have discussed with [redacted] that at this time, I am still unclear as to the exact etiology of [redacted] dizzy spells, as they may represent a migraine variant, or they also may be secondary to these AVM's. As such, I have recommended further evaluation by a neurosurgeon and have referred [redacted] to Dr. [redacted] for further evaluation.

Figure 6-7 *ENT letter I.*

This next letter (Figure 6-8), also from an ENT, provides further details of the complex history. The early history included circulatory problems, namely, mitral valve prolapse, followed by complaints of dizziness. Then, the patient had an episode of vertigo. At that time, a CT scan and MRIs were negative. Then two months ago, the patient began to experience vertigo as well as hearing loss in the left ear, which lasted briefly and resolved spontaneously. Workup to date has included an audiogram, ENG, and MRI. (The interpretations of these three tests were provided in the preceding paragraphs.)

______ is ___year old ______ with a history of mitral valve prolapse, who was referred to me ______ for problems with dizziness. She states that approximately ___ years ago, she had an episode of sudden onset vertigo, which abated over a few days, and evaluation at that time, including CT Scan and MRI, was negative. However, approximately two months ago, she began experiencing intermittent bouts of frank vertigo, as well as hearing loss in the left ear, which lasted approximately 20/minutes at the time and resolved spontaneously. She states that associated with these she does develop some nausea, however, denies any vomiting or loss of consciousness. She does claim to some vague complaints of left arm weakness with these episodes.

Work-up to date has included an audiogram, ENG, and MRI Scan, which I have included for your review. The audiogram was essentially normal. The ENG did reveal some findings suggestive of central vestibular dysfunction. The MRI Scan reveals the documented three small AVM's.

Figure 6-8 *ENT letter II.*

The next document (Figure 6-9) is a letter from a consulting neurosurgeon. This patient's history includes unusual neurological problems dating back approximately 20 years. More recently, about two years ago, the patient began to have unusual spells of passing out. These occurred about eight times during a year and then spontaneously stopped. Approximately one year ago, other unusual spells started, where the patient would feel dizzy and lightheaded while speaking and then would black out for a few seconds. A cardiological exam revealed mitral valve prolapse. An EEG (electroencephalogram) was negative. More recently, the patient developed new symptoms, including the sudden onset of feeling she was being pushed to the right side. This would be accompanied by a spinning sensation. She had a series of such attacks, with subsequent attacks lasting longer.

A more recent neurosurgical evaluation (Figure 6-10) provides some closure on the overall assessment. This neurosurgeon states that this woman had been his patient previously, "for exactly the same problems she is now having." They did a thorough workup of the patient in the past, and "nothing unusual was found." This included evoked responses that were unremarkable

______ has been in good health except for some unusual neurological problems dating back to the late 1970's.

I will first begin with recent complaints which started about two years ago. ___ began having some type of unusual passing out spells. In total she had about seven or eight spells until one year ago when they ceased, but were replaced by other unusual spells.

The spells of blackening out occurred without warning. She would be speaking or sitting or standing when all of a sudden she would become dizzy and lightheaded and then would usually blackout. It would only be seconds. It was never accompanied by tongue bite incontinency or jerking movements of the extremities. She would make a quick recovery although she would still feel lightheaded after the episode for a few minutes.

She had a cardiologic work up at that time which revealed mitral valve prolapse. An electroencephalogram was negative.

She was started on some medication and evidentally she did well until she developed these new episodes in ___ Again, without warning, she would feel like she was being pushed to the right side. This would be accompanied by a spinning sensation, but no true headache.

She has had numerous such attacks, each one lasting somewhat longer than the previous one. That is the reason for her evaluation today

Figure 6-9 *Neurosurgeon letter I.*

and MRI studies of the brainstem, all of which were normal. Also, the patient previously had a complete cardiac workup and an ENT workup, too. An MRI study demonstrated suspected small vascular malformations, but the arteriogram did not demonstrate any abnormality. In addition, the neurosurgeon evaluated the patient on the date of this letter and found that the patient "is still intact." He notes that she was seen previously by a number of physicians with different specialties, all of whom found nothing unusual. He then states, "She has never had a history of documented neurological disease." In fact, when the neurosurgeon has seen the patient in the past, "everything has been normal." Thus, to interpret his statements, he is saying that extensive testing and examination have found nothing wrong with this patient. Thus, the complaints appear to be without an objective identifiable basis. This may reflect a number of possibilities, including an emotional component, malingering, and/or an organic cause that is not yet detectable using available technology.

I appreciate you referring ________ back to me.

She has been a patient of mine in the past and she has had a complete work up for exactly the same problem she is now having which consists of intermittent vertiginous spells which she has several times a week lasting various periods of time, anywhere from seconds to hours.

She has had these in the past and, in fact, we did a very thorough work up in ____ because of similar problems at which time nothing unusual was found. We had even done MRI studies of her brainstem at the time and these were all normal. Evoked responses were also unremarkable.

She has undergone a complete cardiac work up in the past and most recently has undergone ENT work up by Dr. ________ and then a complete work up by you. To briefly review what has been done, she underwent studies by Dr. ________ all of which were normal. She was referred to you. An MRI study demonstrated, as you know, suspected small vascular malformations. Arteriograms did not demonstrate any abnormality.

I have evaluated her today and I should tell you that from a neurological standpoint she is still intact and yet I do wish to tell you some other points of interest.

Back in the 's she underwent a complete work up for headaches and was seen by another neurologist, also by Dr. ____ and by Dr. ______ all of whom found nothing unusual.

She had never had a history of documented neurological disease and, in fact, when I have seen her in the past everything has been normal.

Figure 6-10 *Neurosurgical letter II.*

Case Study 5

This police accident report (PAR) (Figure 6-11) narrative describes a pedestrian receiving a right leg injury as the result of being struck by a car. The complaint may be interpreted as describing an impact that was not accidental.

The next document (Figure 6-12), an emergency department (ED) record, describes a patient who has tenderness of the distal third of the thigh, that is, above the knee. This would be consistent with the pedestrian impact scenario. Although the thigh injury could be due to the pedestrian-ground impact, with the limited information provided, it seems more likely to be due to a vehicle-pedestrian impact, with the front of the vehicle striking the pedestrian slightly above the knee.

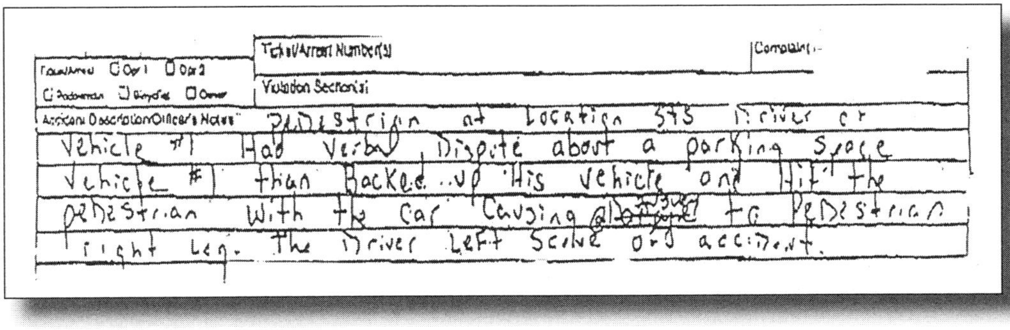

Ticket/Arrest Number(s) | Complaint
Fatal/Injured ☐ Op 1 ☐ Op 2
☐ Pedestrian ☐ Bicyclist ☐ Owner | Violation Section(s)
Accident Description/Officer's Notes: Pedestrian at location states Driver of Vehicle #1 Had verbal Dispute about a parking Space Vehicle #1 than Backed up His vehicle and Hit the pedestrian with the car causing [illegible] to Pedestrian right leg. The Driver Left Scene of accident.

Figure 6-11 *Police accident report of a pedestrian struck by a car.*

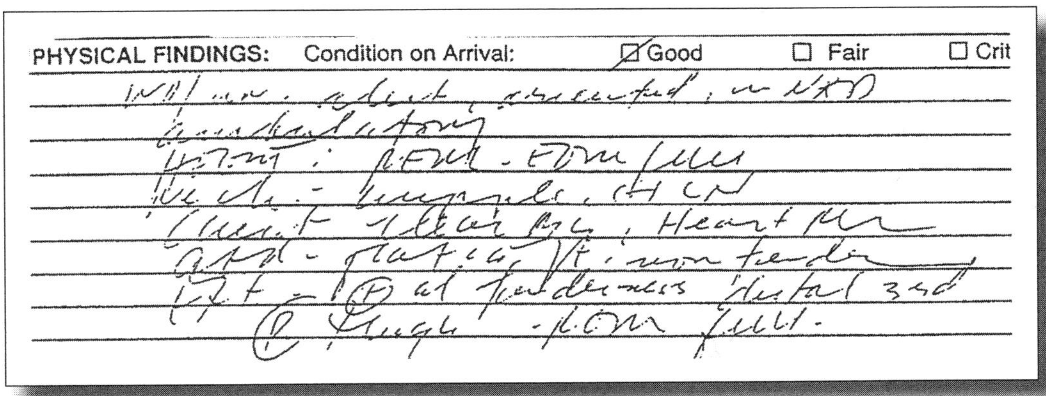

PHYSICAL FINDINGS: Condition on Arrival: ☒ Good ☐ Fair ☐ Crit

Figure 6-12 *Emergency room record of a pedestrian impact.*

Case Study 6

This case study utilizes MRI and electrodiagnostic testing that was conducted in response to patient complaints regarding the right upper limb. In his letter (Figure 6-13) to the physician that asked him to consult, the radiologist provides the impression that there is "C3–C4 through C6–C7 bulging discs with cord flattening." Thus, the protruding intervertebral discs are not only contacting the spinal cord, but pushing on it sufficiently to cause some cord deformation (flattening.) It would not be surprising if this amount of compression produced some clinical manifestations; indeed, the next records (Figures 6-14 through 6-16) report on electrodiagnostic studies designed to assess the electrical conducting properties of the nerves, in conjunction with patient complaints of neck pain radiating to the right upper extremity with numbness and paresthesias. Of particular interest is the F/H report that presents conduction

IMPRESSION: C3-C4 through C6-C7 bulging discs with cord flattening is noted.

Figure 6-13 *Cervical MRI, cord flattening.*

Chief Complaint: Neck pain radiating to the right upper extremity with numbness and paraesthe

Nerve Conduction Report:

Motor Nerves

Nerve	Site	Onset Lat (ms)	Amplitude	Area (mVms)	Duration (ms)	Seg Name	Delta (ms)	Distance (cm)	Velocity (m/s)
L Median	APB		O-P (mV)	Neg	Neg		O		
	Wrist	3.47	9.62	38.630	6.52	Elbow-Wrist	4.22	26.00	61.6
	Elbow	7.69	9.52	38.455	6.61				
R Median	APB		O-P (mV)	Neg	Neg		O		
	Wrist	3.52	10.70	33.085	5.30	Elbow-Wrist	4.27	26.00	61.0
	Elbow	7.78	10.32	32.034	5.30				
L Ulnar	ADM		O-P (mV)	Neg	Neg		O		
	Wrist	3.05	10.31	34.320	6.66	Elbow-Wrist	4.50	26.00	57.8
	Elbow	7.55	9.15	31.275	6.94				
R Ulnar	ADM		O-P (mV)	Neg	Neg		O		
	Wrist	3.66	9.01	30.630	5.53	Elbow-Wrist	4.36	26.00	59.6
	Elbow	8.02	9.23	29.860	5.77				

Sensory Nerves

Nerve	Site	Onset Lat (ms)	Peak Lat (ms)	Amplitude	Area (mVms)	Duration (ms)	Seg Name	Delta (ms)	Distance (cm)	Velocity (m/s)
L Median 2	2ndDig			P-T (μV)	Neg	Neg		P		
	Wrist	3.06	3.84	40.69	0.033	1.75	Wrist-2ndDig	3.84	16.00	41.6
R Median 2	2ndDig			P-T (μV)	Neg	Neg		P		
	Wrist	3.31	3.94	53.93	0.012	1.06	Wrist-2ndDig	3.94	16.00	40.6
L Ulnar	5thDig			P-T (μV)	Neg	Neg		P		
	Wrist	2.91	3.69	32.17	0.016	1.38	Wrist-5thDig	3.69	16.00	43.4
R Ulnar	5thDig			P-T (μV)	Neg	Neg		P		
	Wrist	2.81	3.66	47.39	0.022	1.41	Wrist-5thDig	3.66	16.00	43.8

F/H Report:

Nerve	Muscle	Lat1 (ms)	Lat2 (ms)	Lat2 - Lat1 (ms)	Amplitude (μV)
L Median F	APB	3.13	27.50	24.38	
R Median F	APB	3.91	27.73	23.83	
L Ulnar F	ADM	3.83	28.98	25.16	
R Ulnar F	ADM	3.36	30.70	27.34	

EMG Report:

Side	Muscle	Nerve	Root	INS	FIBS	PSW	AMP	DUR	POLY	REC	IP	COMMENT
L	CervParaC6-7			Nml	0	0	Nml	Nml	Nml	Normal	Nml	
R	CervParaC6-7			Inc	1+	2+	Nml	Nml	Nml	Normal	Nml	
L/R	Biceps	Musc	C5-6	Nml	0	0	Nml	Nml	Nml	Normal	Nml	

Figure 6-14 *Upper extremity nerve conduction report.*

F/H Report:

Nerve	Muscle	Lat1 (ms)	Lat2 (ms)	Lat2 - Lat1 (ms)	Amplitude (μV)
L Median F	APB	3.13	27.50	24.38	
R Median F	APB	3.91	27.73	23.83	
L Ulnar F	ADM	3.83	28.98	25.16	
R Ulnar F	ADM	3.36	30.70	27.34	

Figure 6-15 *Upper extremity nerve conduction report—close-up.*

Impression:
The above electrodiagnostic findings are consistent with right nerve root dysfunction at the level of C6-C7. Prolonged Right Ulnar F-wave latency is evident. Clinical follow up is suggested.

Figure 6-16 *Upper extremity nerve conduction report impression.*

times for the left and right median and ulnar nerves. (The name of this report is derived from the so-called F response, originally measured for the foot, and the H reflex, named after the clinician Hoffmann who first evoked the response [Preston, 1998].) The distances for the left and right sides are approximately equal; therefore, the conduction times also may be considered as representing velocities. Looking in the column labeled "Lat2 – Lat1 (ms)," we note that the times for the left and right median nerves differ by only 24.38 – 23.83 = 0.55 ms, a difference that is within normal variability. However, when we compare the latency (time) for the left and right ulnar nerves, we observe a difference of 27.34 – 25.16 = 2.18 ms, with the right ulnar nerve displaying an abnormally slow time (i.e., an abnormally slow conducting velocity). This leads the clinician to report an impression that these findings are "consistent with right nerve root dysfunction at the level of C6–C7. Prolonged Right Ulnar F-wave latency is evident." Thus, the MRI and nerve conduction study (NCS) present a consistent picture of a bulging disc pressing on and compressing a nerve root and interfering with the nerve's functioning, manifesting as pain at the site of compression (the neck) and sensory deficits (tingling, paresthesias) at the structure being innervated.

Case Study 7

This case study, similar to the previous one, also combines MRI and electrodiagnostic studies but this time for the lower extremity. The patient's complaints are low back pain radiating to the right lower extremity, with numbness and parasthesias.

In his letter (Figure 6-17) to the referring physician, the radiologist relates the impression that there is "right paracentral herniation at L5-S1, compression of the dura and the right S1 nerve root." Thus, the disc material that herniated is pushing backward on the right side and compressing the dura mater (the protective membrane) and the right nerve root, the branch of the spinal cord. It would not be surprising if this amount of compression produced some clinical manifestations; indeed, the next records (Figures 6-18 and 6-19) report on electrodiagnostic studies designed to assess the electrical conducting properties of the nerves, in conjunction with the patient's complaints of low back pain radiating to the right lower extremity with numbness and paresthesias.

IMPRESSION:

There is a right paracentral herniation at L5-S1, compression of the dura and the right S1 nerve root.

Figure 6-17 *L/S (lumbar spine) MRI, dura and nerve root compression.*

F/H Report:

Nerve	Muscle	Lat1 (ms)	Lat2 (ms)	Lat2 - Lat1 (ms)	Amplitude (µV)
L Default H		5.63	31.09	25.47	
R Default H		5.31	31.72	26.41	
L Peroneal F		5.00	49.38	44.38	
R Peroneal F		4.53	52.97	48.44	
L Tibial F		4.69	49.38	44.69	
R Tibial F		4.53	49.06	44.53	

Figure 6-18 *LE (lower extremity) NCR (nerve conduction report).*

Impression:
The above electrodiagnostic findings are consistent with right nerve root dysfunction at the level of L4-L5.Prolonged Right Peroneal F-wave latency is supportive of the diagnosis. Clinical follow up is suggested.

Figure 6-19 *Lower extremity (LE) nerve conduction report (NCR).*

Of particular interest is the F/H report that presents conduction times for the left and right peroneal and tibial nerves. (Again, the name of this report is derived from the so-called F response, originally measured for the foot, and the H reflex, named after the clinician Hoffmann who first described evoking the response [Preston, 1998].) The distances for the left and right sides are approximately equal; therefore, the conduction times also may be considered as representing velocities. Looking in the column labeled "Lat2 – Lat1 (ms)," we note that the times for the left and right tibial nerves differ by only 44.69 – 44.53 = 0.16 ms, a difference that is within normal variability. However, when we compare the latency (time) for the left and right peroneal nerves, we observe a difference of 48.44 – 44.38 = 4.06 ms, with the right peroneal nerve displaying an abnormally slow time (i.e., an abnormally slow conducting velocity).

Although the MRI report identifies compression of the right S1 nerve root while the NCV points to compression at the right L4/L5 nerve root, the clinician apparently is satisfied that these results and the clinical information are consistent. He reports (Figure 6-19) an impression that these findings are "consistent with right nerve root dysfunction at the level of L4-L5. Prolonged Right Peroneal F-wave latency is supportive of the diagnosis." Presumably, the consistent picture is one of material from a ruptured disc pressing on and compressing a nerve root and that compression in turn interfering with the nerve's functioning and manifesting as pain at the site of compression (lower back) and sensory deficits (tingling, paresthesias) at the structure being innervated.

Case Study 8

This example includes a police accident report (PAR) (Figure 6-20) as well as some medical records (Figures 6-21 through 6-25).

The MVA was somewhat complex, in that after an impact involving the right rear of a truck and the front of a passenger car, the two vehicles became "stuck" together. Then the passenger car was dragged along for some distance by the truck before the two vehicles separated. The truck driver then left the accident scene. Thus, the impact scenario for the passenger vehicle was somewhat of a frontal impact, with an additional component of a sideways dragging.

114 ACCIDENT DESCRIPTION
DRIVER #1 STATED, "I SPUN OUT AND WAS HIT BY THE TRUCK, WHO THEN
DRAGGED ME FOR A SHORT DISTANCE AND THEN LEFT." INVESTIGATION AT
THE SCENE REVEALED THAT VEH #1 WAS TRAVELING IN THE RIGHT
LANE WHEN HE LOST CONTROL ON THE SNOW COVERED ROAD AND SPUN
INTO THE CENTER LANE, AND STRUCK THE REAR SIDE OF A TT. VEH #2
THEN DRAGGED VEH #1 UNDER IT'S TIRES BEFORE FLEEING THE SCENE.

Figure 6-20 *Police accident report of a dragged vehicle.*

AP, LATERAL VIEWS OF THE CERVICAL SPINE: There is some straightening of the usual cervical lordosis, however, vertebral body heights and interspacing appear intact with good alignment. No fractures are identified.

IMPRESSION: Straightened curve may be indicative of muscle spasm or be positional.

Figure 6-21 *Impression of the cervical spine.*

EXAMINATION PERFORMED: MRI BRAIN W/O & W CONTRAST

CLINICAL INFORMATION: Headaches.

IMPRESSION: MRI BRAIN - Normal.

Figure 6-22 *A head MRI of the brain.*

MRI OF THE LUMBAR AND CERVICAL SPINE

IMPRESSION:

1. Normal MRI of the lumbar spine.
2. Normal MRI of the cervical spine.

Figure 6-23 *An MRI of the cervical and lumbar spines.*

IMPRESSION:

Normal EEG in the awake state.

Figure 6-24 *An EEG impression.*

His EEG is normal, and his MRI of the brain is normal

DIAGNOSIS: Tension headache.

Figure 6-25 *Physician's office note.*

The driver of the passenger vehicle complained of headache and, as a result, underwent a series of radiologic procedures aimed at determining the cause of the headache. A series of neck x-rays (Figure 6-21) did not show any fractures and showed good vertebral alignment. However, the radiologist's impression was that the spine appeared more straight than typical and that this straightening may have been "indicative of muscle spasm or be positional." That is, the spine in the neck region, when viewed from the side, normally has a curved appearance (concave posteriorally), but if something causes this curve to be reduced, it "shows up" on imaging as a spine with a straighter appearance. Cervical straightening may or may not be indicative of pathology. Some spines are straighter than others, and the particular spine simply may be naturally straighter than average, may have been straightened by muscular forces, or may merely reflect the neck being straightened at the time of imaging (by the head being in a nodding position, with the chin close to the chest). Thus, the radiologist has the impression that in this instance, the straightening may or may not indicate muscle spasm. (A lack of neck injury would also be an indicator that the head did not move violently during this motor vehicle accident [MVA]).

Next, a series of MRIs were conducted. A head MRI (Figure 6-22) was performed and was interpreted as being normal. The MRIs of the cervical and lumbar spine (Figure 6-23) also were interpreted as being normal. Finally, an EEG (electroencephalogram) (Figure 6-24) was performed to examine electrical activity in the brain, and this was read as being normal, too. Thus, these tests ruled out different possibilites, and the physician's office note (Figure 6-25) indicated the remaining diagnosis, namely, "tension headache."

Chapter 7

א

Advanced Case Study I: Pre-Existing Conditions

Introduction

These case studies differ from the preceding cases in that these have a greater variety of records. One of the challenges is to integrate the various records, which sometimes may appear to give contradictory information, into a cohesive whole. Working through the cases in this section provides the following: (1) additional practice in reading and interpreting medical records, (2) an opportunity to further develop critical reasoning skills, and (3) in conjunction with the detailed index, serves as a reference for a variety of records and injury scenarios. For easy reference, the case studies will be organized one per chapter.

Introduction

As mentioned in the preceding paragraph, this case study provides a number of records (Figures 7-1 through 7-19). The goal is to review the records and integrate them to form working hypotheses regarding identifying any significant injuries and their cause(s). The records are arranged in five groups, designated as follows: emergency room (ER) (Figures 7-1 through 7-6), chiropractor I (x-rays) (Figures 7-7 through 7-9), bone scan (Figure 7-10), radiology consultation (Figure 7-11), chiropractor II (Figure 7-12), independent chiropractic exam (Figures 7-13 through 7-17), and magnetic resonance imaging (MRI) (Figures 7-18 and 7-19). An additional document, a chronology or time line, has been included (Figure 7-20). Especially in cases such as this, where the history includes pre-existing conditions and/or more than one trauma and/or treatment over a period of time, it often is useful to create a time line or chronology. It may be useful to create separate chronologies for events (e.g., motor vehicle accident [MVA]) and treatments or, as was done in this case, to combine the two into a single chronology. The medical records, as presented, do not include dates; thus, the chronology also helps to make the example more realistic, in terms of the information that would be available.

Note: A discussion follows Figure 7-20. The 20 figures will be repeated in the Discussion section, with callouts and/or marginalia added. For those readers who wish to try the analysis on their own first, it is recommended that they not look past Figure 7-20 until they have tried the exercise. For those who prefer another "guided tour," it is recommended they go directly to the Discussion section and the second set of figures.

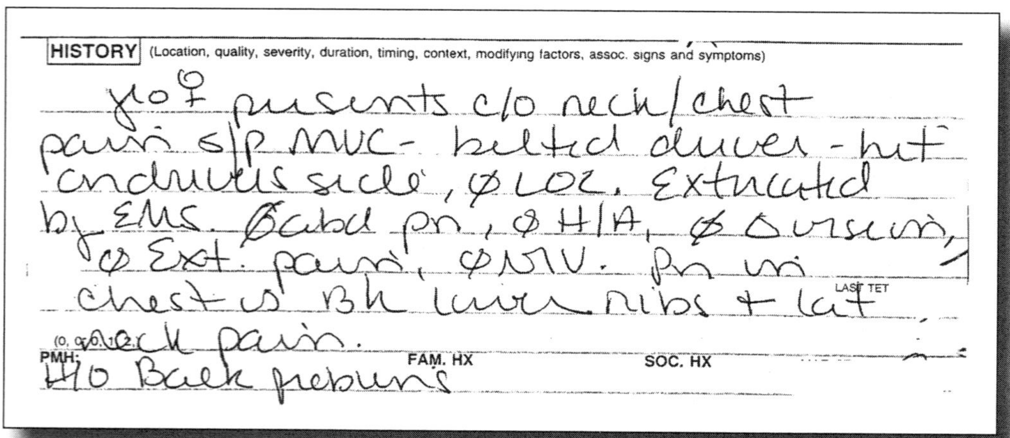

Figure 7-1 ER history I.

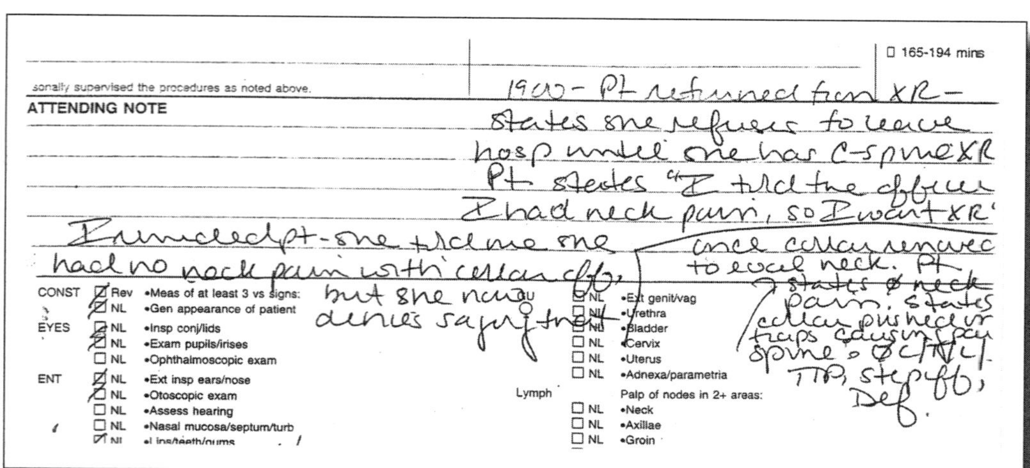

Figure 7-2 ER history II.

PART I - PRELIMINARY READING

ATTENTION: THIS IS A PRELIMINARY READING BY A RADIOLOGY RESIDENT AND WILL BE REVIEWED BY AN ATTENDING RADIOLOGIST.

Date **Time Read ___:___ am pm**

TYPE OF EXAM:

1) Faint lucency seen projected over spinous process of C-6 this may represent mach effect rather than Fx - since seen only on one swimmer's view

2) otherwise remaining C-spine wnl.

Figure 7-3 *Mach effect.*

IMPRESSION: Questionable hairline fracture of uncertain age involving the tip of the spinous process of C6. Clinical correlation is recommended. Rest of the cervical spine is negative.

Figure 7-4 *ER impression of the cervical spine 1.*

FOUR VIEWS OF THE CERVICAL SPINE AND CROSS TABLE LATERAL VIEW:

CLINICAL INFORMATION: Injury.

There is normal alignment of the cervical vertebral bodies. There is a small possible hairline fracture of uncertain age noted in the tip of the spinous process of C6. The body appears normal. Lamina line is normal. There is congenital fusion of the lamina and pedicle of C2 and C3. The predental space and the atlantoaxial joint appear normal.

Figure 7-5 *ER impression of the cervical spine 2.*

BILATERAL RIBS:

CLINICAL INFORMATION ___-year-old with injury. This is to evaluate for fracture.

There is no evidence of fractures in both rib cages. The lungs are well expanded.

IMPRESSION: Normal rib cage.

Figure 7-6 *ER impression of a rib x-ray.*

TYPE OF X-RAY: AP LATERAL CERVICAL SERIES

1. Shows no evidence of active bony pathology or acute bony fracture.
2. The spine appears midline.
3. Unco vertebral joints appear well maintained.
4. The cervical lordosis is well maintained.
5. Disc intervals appear well maintained.
6. There is a very mild facet imbrication in the mid axial region of the cervical spine.
7. There appears to be C2/3 arthrodeses nature fusion, which could be congenital.

Figure 7-7 *Chiropractor report I—cervical.*

TYPE OF X-RAY: AP LATERAL THORACIC VIEW

1. There is some compression in the AP projection of the T3 segment which should be evaluated with nuclear bone scan due to recent auto accident otherwise the thoracic kyphosis appears within normal limits.
2. Disc intervals appear well maintained.
3. There is no evidence of osteoarthrosis.
4. All lamina pedicle junctions are viewable in the AP projection.
5. All rib interspaces are well maintained and symmetrical.

Figure 7-8 *Chiropractor report I—thoracic.*

TYPE OF X-RAY: AP LATERAL LUMBO-PELVIC VIEW

1. Shows no evidence of active bony pathology or acute bony fracture.
2. All lamina pedicle junctions are viewable in the AP projection.
3. There are only 4 lumbar vertebrae.
4. The lumbar lordosis is well maintained.
5. The lumbar lordosis is well maintained.
6. Disc intervals are well maintained.
7. There is apparent left leg length deficiency with a left PI ilia malalignment.
8. Hip joints are bilaterally symmetrical and well maintained.
9. Sacroiliac joints are bilaterally symmetrical and well maintained.
10. There appears to be some facet arthrosis in the lumbosacral facet.
11. The coccyx is to the left of the symphysis pubis.

Figure 7-9 *Chiropractor report I—lumbar/sacral.*

IMPRESSION: MULTIPLE FOCI OF ABNORMAL INCREASED UPTAKE IN THE LEFT AND RIGHT RIBS COMPATIBLE WITH ACUTE FRACTURES. INCREASED UPTAKE L5 BILATERALLY MAY BE DUE TO ARTHRITIS. THE DISTRIBUTION IS NOT TYPICAL FOR A COMPRESSION FRACTURE. ABNORMAL INCREASED UPTAKE IN THE MANUBRIUM MOST COMPATIBLE WITH ACUTE FRACTURE.

Figure 7-10 *Impression of the bone scan.*

BILATERAL RIBS, CHEST AND STERNUM

Clinical data: Status post MVA in ________ and previous radionuclide bone scan showed fractures of multiple ribs and sternum.

Findings: AP and both oblique views of the ribs and a frontal chest radiograph were obtained. Lateral and AP views of the sternum were also obtained. Multiple views of the bilateral ribs and sternum show evidence of a healing fracture involving the posterior aspect of the left sixth, seventh, eighth and ninth ribs as well as the lateral aspect of the left sixth and seventh ribs with callous formation but no significant displacement or angulation of the fracture fragments can be seen. The lateral projection of the sternum shows minimal cortical irregularity of the manubrium indicating a healing fracture but no significant displacement is seen on the lateral projection. If indicated, CT scan would be more sensitive in evaluating the sternum.

PA view of the chest shows the lungs to be clear. The heart size, pulmonary vascularity and mediastinum are within normal limits. There is no pleural effusion or pneumothorax. The left rib fractures are not well seen on the PA chest.

IMPRESSION: HEALING FRACTURES OF THE LEFT SIXTH, SEVENTH, EIGHTH AND NINTH RIBS AS WELL AS THE MANUBRIUM. NO SIGNIFICANT DISPLACEMENT OF THE FRACTURE FRAGMENTS CAN BE SEEN.

Thank you for your kind referral.

Figure 7-11 *Radiology consultation I.*

Past History:
Past medical history includes that of neck pain, midback pain, and sacro-iliac pain due to a motor vehicle accident on ______ where she was the passenger . Chiropractic care was initiated with Dr. ______ for a period of 6 months during which time she made a full and complete recovery. Ms. ____ notes no other major medical history, and is an otherwise healthy ____ year old female.

Figure 7-12 *Chiropractor report II.*

To Whom it May Concern,

As per your request, an independent chiropractic examination was performed on the above captioned claimant at ______ Chiropractic on . ______ The findings of that examination are presented as follows.

Figure 7-13 *Independent chiropractic exam (ICE) I.*

Past history:

The claimant says he has no previous surgical history. She denies prior illnesses. She had a prior occupational injury in ____ when she worked for ________ when she jumped down from a machine and slipped on a popsicle on the floor, hitting her back on the stairs. She tells me she injured her neck and back during that accident and bruised her lung. She treated with Dr. ______ , a Chiropractor at that time. She says she was off work for 4 and a half years as a result of that accident.

She has a prior motor vehicle accident in ______ when she was the back seat passenger in a car, the vehicle was rear ended by another driver. She was taken to a hospital in _____ She treated with a Chiropractor, Dr. in ______ from _____ She has had no follow up care since that time regarding that accident.

Figure 7-14 *Independent chiropractic exam (ICE) II.*

Impression: Cervical, thoracic and lumbar sprain strain syndrome, spasm, joint dysfunction, possible fractured ribs. Possible manubrium fracture.

Figure 7-15 *Independent chiropractic exam (ICE) III regarding rib fracture.*

Need for diagnostic testing: I cannot rule out the need for further diagnostic testing at this time, especially in the form of plain film x-rays of the thoracic spine, ribs and manubrium to rule out (o rather verify, rule in) healing status of fractures. I cannot rule out the need for additional advance diagnostic testing at this time, but cannot recommend it at this time either.

Disability: The claimant was not employed at the time of the accident, thus disability status is not an issue. However, I do not identify any element of disability on today's examination that would stop her from seeking employment.

Figure 7-16 *Independent chiropractic exam (ICE) IV regarding disability.*

Causal Relationship: With the exclusion of suspected fractures, I find causal relationship to the accident of ______ and her current symptoms. While it is very possible (and probable) that she may have sustained fractures during this accident, all plain film testing does not verify this. Bone scan is sensitive to fracture, but not specific to fracture, thus, until I can see proof otherwise, I cannot find causal relationship to fracture and the ______ accident.

Figure 7-17 *Independent chiropractic exam (ICE) V regarding causality.*

<u>IMPRESSION:</u> MILD DIFFUSE POSTERIOR BULGE OF THE ANNULUS FIBROSUS AT C2-3.

SMALL TO MODERATE SIZE CENTRAL DISC PROTRUSION AT C3-4.

VERY SMALL C4-5 CENTRAL DISC PROTRUSION WITHOUT MASS EFFECT UPON THE CERVICAL CORD OR NEURAL FORAMINAL COMPROMISE.

MODERATE SEVERITY CIRCUMFERENTIAL BULGE OF THE ANNULUS FIBROSUS AT C5-6.

MODERATE SEVERITY RIGHT PARACENTRAL FOCAL DISC PROTRUSION AT C6-7.

Thank you for your kind referral.

Figure 7-18 *Impression of the cervical MRI.*

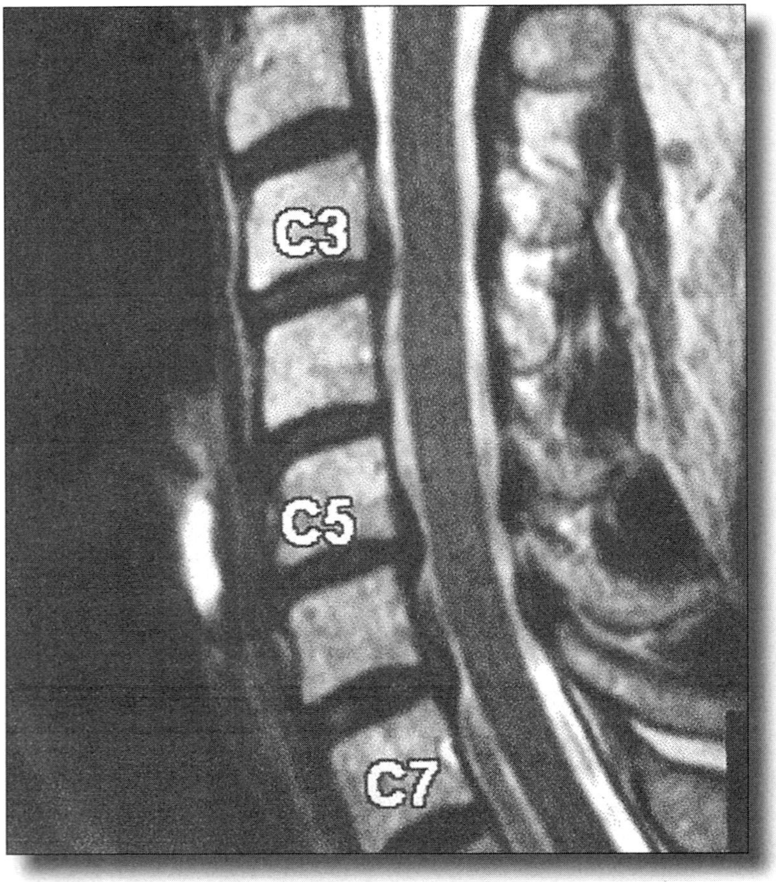

Figure 7-19 Cervical MRI image.

Time Line (Events & Treatment)

0	Congenital	Partial fusion - C2/3, only 4 lumbar vert.
1	1/1/1990	Occupational injury, off work 4-1/2 yrs.
2	1/1/2000	mva, rear-ended, treated for 6 mos.
3	1/1/2002	Subject MVA, extract & take to ER, XR's
4	1/9/2002	DC1 OV, C/T/L XR's
5	1/13/2002	Bone Scan, Chest XR's
6	2/17/2002	DC2 OV
7	4/9/2002	IME
8	5/27/2002	Neck MRI

Figure 7-20 Time line.

Note: Immediately following this note, there is a discussion of the case, followed by a repeat of the preceding 20 figures. However, this second version of the figures includes callouts and/or marginalia. There are a total of 20 images, with the "original" figures designated Figures 7-1 through 7-20, and the annotated set of figures designated Figures 7-21 through 7-40. Figure 7-21 is an annotated version of Figure 7-1, Figure 7-22 is an annotated version of Figure 7-2, and so on.

Again, for those readers who wish to try the analysis on their own first, it is recommended that they not read the Discussion until they have tried the exercise. For those who prefer another "guided tour," it is recommended they go directly to the Discussion (immediately following this paragraph) and refer to the second set of figures.

Discussion

This ER record (Figure 7-21) provides some basic details regarding the history of the event and patient, including that she was the seat-belted driver of a vehicle that was impacted on the driver's side, that she complained of chest and neck pain, and that she had a history of previous back complaints. This record also states that the patient was "extricated by EMS." Thus, the emergency medical services (EMS) report may contain objective firsthand information regarding seatbelt usage and so forth. The ER record (Figure 7-22) also provides a somewhat unusual notation regarding the patient refusing to leave the hospital until she had a cervical spine x-ray. There also is a notation that she told the nurse her neck did not hurt and then denied saying that. It appears that this record reflects an emotional overlay and, at the least, indicates some friction between the nurse and the patient, which may call into question the objectivity of this record.

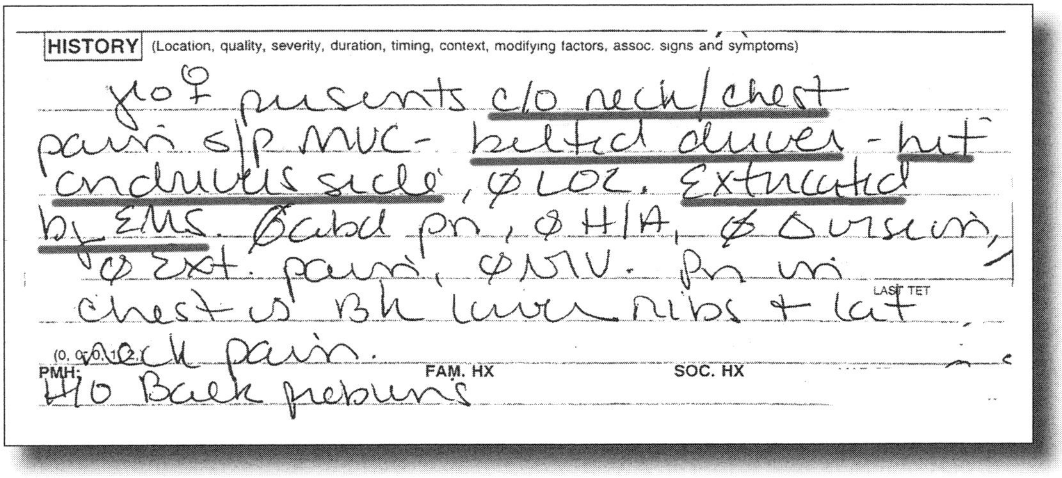
HISTORY (Location, quality, severity, duration, timing, context, modifying factors, assoc. signs and symptoms)

40 ♀ presents c/o neck/chest pain s/p MVC - belted driver - hit on drivers side, ∅ LOC. Extricated by EMS. Back pn, ∅ H/A, ∅ dizziness, ∅ ext. pain, ∅ N/V. Pn in chest w/ Bh lower ribs + lat neck pain.

LAST TET

PMH: H/O Back problems

FAM. HX

SOC. HX

Figure 7-21 *ER history I.*

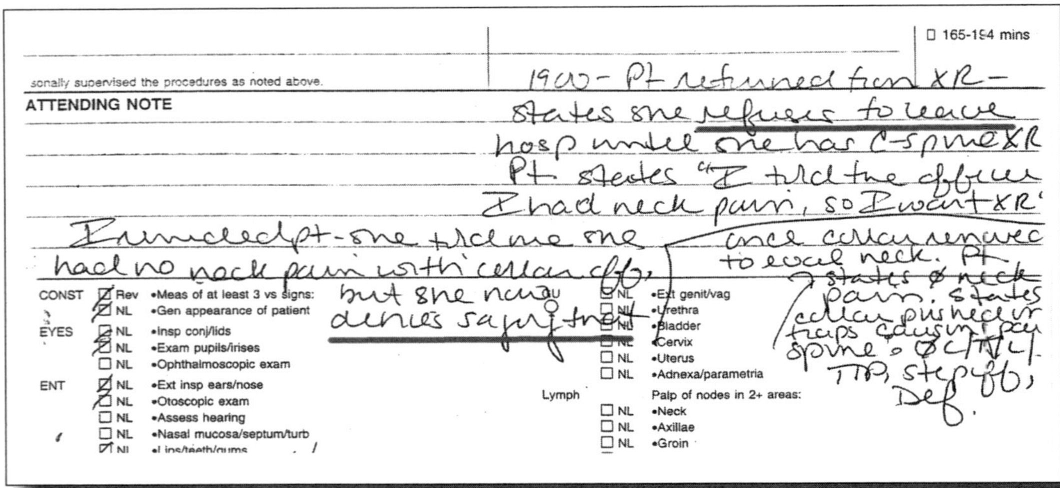
☐ 165-184 mins

sonally supervised the procedures as noted above.

ATTENDING NOTE

1900 - Pt returned from XR - states she refuses to leave hosp until she has C-spine XR. Pt states "I told the officer I had neck pain, so I want XR'

Examined pt - she told me she had no neck pain with collar off, but she now denies saying that

once collar removed to eval neck. Pt states ø neck pain. States collar pushed on traps causing pain. spine ø CTNL. TTP, step off, Def.

CONST ☑ Rev •Meas of at least 3 vs signs:
☑ NL •Gen appearance of patient
EYES ☑ NL •Insp conj/lids
☑ NL •Exam pupils/irises
☐ NL •Ophthalmoscopic exam
ENT ☑ NL •Ext insp ears/nose
☑ NL •Otoscopic exam
☐ NL •Assess hearing
☐ NL •Nasal mucosa/septum/turb
☑ NL •Lips/teeth/gums

☑ NL •Ext genit/vag
☑ NL •Urethra
☐ NL •Bladder
☐ NL •Cervix
☐ NL •Uterus
☐ NL •Adnexa/parametria
Lymph Palp of nodes in 2+ areas:
☐ NL •Neck
☐ NL •Axillae
☐ NL •Groin

Figure 7-22 *ER history II.*

The next few records (Figures 7-23 through 7-26) are from the Imaging Services Department. The preliminary radiology report (Figure 7-24), prepared by a medical resident, notes a line on C6 that could represent a spinous process fracture. However, the resident discusses that it may be an imaging artifact called the Mach effect (Harris, 1996). The attending radiologist reflects this in the radiology report, which notes a "small possible hairline fracture of uncertain age" on the tip of the C6 process (Figure 7-24). The report also notes congenital fusion of the lamina and pedicles of C2 and C3 (Figure 7-25). Thus, parts of the second and third cervical vertebrae are fused together, and the radiologist is opining that this structural anomaly was developmental and not attributable to this trauma. Note that this lack of symmetry and mobility of the cervical spine may affect its ability to withstand axial and/or rotational loading and may affect its susceptibility to degenerative disease.

The next record from the Imaging Services Department (Figure 7-26) reports on the results of rib x-rays that were taken to rule out rib fracture. The radiologist did exactly that—he wrote that there was "no evidence of fracture in both rib cages," and his impression was "normal rib cage." This absence of any rib fracture would be one indicator that there was not a significant chest impact. Is this impression consistent with the other imaging reports (e.g., Figure 7-30)?

The next series of documents (Figures 7-27 through 7-29) relates to a series of post-ER x-ray results, summarized by the first of several chiropractors who examined and/or treated the patient. The cervical series notes the C2/C3 fusion as well as a mild facet imbrication (overlapping). The thoracic series notes some compression of the T3 vertebra and that all rib interspaces are well maintained and symmetrical (i.e., no indication of rib fracture). The lumbo-pelvic series notes some structural anomalies, including only four lumbar vertebrae

PART I - PRELIMINARY READING

ATTENTION: THIS IS A PRELIMINARY READING BY A RADIOLOGY RESIDENT AND WILL BE REVIEWED BY AN ATTENDING RADIOLOGIST.

Date **Time Read** ____:____ **am pm**

TYPE OF EXAM:

1) Faint lucency seen projected over spinous process of C-6 this may represent 'mach-effect' rather than Fx - since seen only on one swimmer's view

2) otherwise remaining C-Spine wnl.

Figure 7-23 *Mach effect.*

IMPRESSION: Questionable hairline fracture of uncertain age involving the tip of the spinous process of C6. Clinical correlation is recommended. Rest of the cervical spine is negative.

Figure 7-24 *ER impression of the cervical spine 1.*

FOUR VIEWS OF THE CERVICAL SPINE AND CROSS TABLE LATERAL VIEW:

CLINICAL INFORMATION: Injury.

There is normal alignment of the cervical vertebral bodies. There is a small possible hairline fracture of uncertain age noted in the tip of the spinous process of C6. The body appears normal. Lamina line is normal. There is congenital fusion of the lamina and pedicle of C2 and C3. The predental space and the atlantoaxial joint appear normal.

Figure 7-25 *ER impression of the cervical spine 2.*

BILATERAL RIBS:

CLINICAL INFORMATION: __-year-old with injury. This is to evaluate for fracture.

There is no evidence of fractures in both rib cages. The lungs are well expanded.

IMPRESSION: Normal rib cage.

Figure 7- 26 *ER impression of a rib x-ray.*

TYPE OF X-RAY: AP LATERAL CERVICAL SERIES

1. Shows no evidence of active bony pathology or acute bony fracture.
2. The spine appears midline.
3. Unco vertebral joints appear well maintained.
4. The cervical lordosis is well maintained.
5. Disc intervals appear well maintained.
6. There is a very mild facet imbrication in the mid axial region of the cervical spine.
7. There appears to be C2/3 arthrodeses nature fusion, which could be congenital.

Figure 7-27 *Chiropractor report I—cervical.*

TYPE OF X-RAY: AP LATERAL THORACIC VIEW

1. There is some compression in the AP projection of the T3 segment which should be evaluated with nuclear bone scan due to recent auto accident otherwise the thoracic kyphosis appears within normal limits.
2. Disc intervals appear well maintained.
3. There is no evidence of osteoarthrosis.
4. All lamina pedicle junctions are viewable in the AP projection.
5. All rib interspaces are well maintained and symmetrical.

Figure 7-28 *Chiropractor report I—thoracic.*

TYPE OF X-RAY: AP LATERAL LUMBO-PELVIC VIEW

1. Shows no evidence of active bony pathology or acute bony fracture.
2. All lamina pedicle junctions are viewable in the AP projection.
3. There are only 4 lumbar vertebrae.
4. The lumbar lordosis is well maintained.
5. The lumbar lordosis is well maintained.
6. Disc intervals are well maintained.
7. There is apparent left leg length deficiency with a left PI ilia malalignment.
8. Hip joints are bilaterally symmetrical and well maintained.
9. Sacroiliac joints are bilaterally symmetrical and well maintained.
10. There appears to be some facet arthrosis in the lumbosacral facet.
11. The coccyx is to the left of the symphysis pubis.

Figure 7-29 *Chiropractor report I—lumbar/sacral.*

and arthritic changes at what typically would be designated as the L5/S1 facet joint and displacement of the coccyx to the left. There is some indication that the anomalies have already affected the region's susceptibility to degenerative changes and load-carrying ability.

The bone scan (Figure 7-30) has identified abnormal metabolic activity in the manubrium and bilaterally in the ribs. This finding is compatible with recent fractures in these regions. The bone scan also reports abnormal activity at the L5 vertebra, with the impression that this may be due to arthritis.

IMPRESSION: MULTIPLE FOCI OF ABNORMAL INCREASED UPTAKE IN THE LEFT AND RIGHT RIBS COMPATIBLE WITH ACUTE FRACTURES. INCREASED UPTAKE L5 BILATERALLY MAY BE DUE TO ARTHRITIS. THE DISTRIBUTION IS NOT TYPICAL FOR A COMPRESSION FRACTURE. ABNORMAL INCREASED UPTAKE IN THE MANUBRIUM MOST COMPATIBLE WITH ACUTE FRACTURE.

Figure 7-30 *Impression of the bone scan.*

Next, the radiologist assesses the bone scan in combination with various other imaging (Figure 7-31) and reports the impression that there were, in fact, fractures of the manubrium as well as the left side ribs 6, 7, 8, and 9 and that these fractures all are healing.

BILATERAL RIBS, CHEST AND STERNUM

Clinical data: Status post MVA in ________ and previous radionuclide bone scan showed fractures of multiple ribs and sternum.

Findings: AP and both oblique views of the ribs and a frontal chest radiograph were obtained. Lateral and AP views of the sternum were also obtained. Multiple views of the bilateral ribs and sternum show evidence of a healing fracture involving the posterior aspect of the left sixth, seventh, eighth and ninth ribs as well as the lateral aspect of the left sixth and seventh ribs with callous formation but no significant displacement or angulation of the fracture fragments can be seen. The lateral projection of the sternum shows minimal cortical irregularity of the manubrium indicating a healing fracture but no significant displacement is seen on the lateral projection. If indicated, CT scan would be more sensitive in evaluating the sternum.

PA view of the chest shows the lungs to be clear. The heart size, pulmonary vascularity and mediastinum are within normal limits. There is no pleural effusion or pneumothorax. The left rib fractures are not well seen on the PA chest.

IMPRESSION: HEALING FRACTURES OF THE LEFT SIXTH, SEVENTH, EIGHTH AND NINTH RIBS AS WELL AS THE MANUBRIUM. NO SIGNIFICANT DISPLACEMENT OF THE FRACTURE FRAGMENTS CAN BE SEEN.

Thank you for your kind referral.

Figure 7-31 *Radiology consultation I.*

The next document (Figure 7-32) is a letter report from the second treating chiropractor. The history related in that document includes a previous MVA when the patient was a passenger and complained of neck, mid-back, and low back pain, for which she received six months of chiropractic treatment. She reports no other major medical history. With regard to the most recent MVA, this history indicates a two-stage impact—first, a vehicle impacted the driver's side, and then another vehicle rear-ended the patient's vehicle. The patient relates that she did not lose consciousness.

The next document (Figures 7-33 through 7-37) is excerpted from the report of an independent chiropractic examination (ICE) by a third chiropractor. Typically, ICEs (or more generally, IMEs [independent medical examinations]) are conducted by non-treating clinicians for the

Past History:
Past medical history includes that of neck pain, midback pain, and sacro-iliac pain due to a motor vehicle accident on ______ where she was the passenger . Chiropractic care was initiated with Dr. ______ for a period of 6 months during which time she made a full and complete recovery. Ms. ____ notes no other major medical history, and is an otherwise healthy ___ year old female.

Figure 7-32 *Chiropractor report II.*

purpose of reviewing treatment, assessing disability, and planning future treatment. In contrast to the history reported by the second chiropractor, the ICE reports a history that includes previous injury to the patient's neck, back, and lung. As a result of these injuries, she was off work for 4-1/2 years. The ICE history also reports on the previous MVA and provides the additional detail that the patient was a rear-seat passenger in a vehicle that was rear-ended, and, as a result, she received chiropractic treatment for six months (Figure 7-34). Regarding the most recent MVA, it is described as first a vehicle impacted at the driver door and then the vehicle was in a rear impact with another vehicle. The patient reports that she was in and out of consciousness. The ICE reports the impression of cervical, thoracic, and lumbar sprain strain syndrome, spasm, joint dysfunction, and possible rib and manubrium fracture (Figure 7-35). Compare this impression of possible fracture to other imaging reports (e.g., Figure 7-30). The ICE finds a causal relationship between the latest MVA and all of the findings, with the possible exception of the rib and manubrium fractures (Figure 7-37). (It is unclear if the clinician conducting the ICE had the latest series of rib and manubrium x-rays.)

To Whom it May Concern,

As per your request, an independent chiropractic examination was performed on the above captioned claimant at ______ Chiropractic on . ______ The findings of that examination are presented as follows.

Figure 7-33 *Independent chiropractic exam (ICE) I.*

Past history:

The claimant says he has no previous surgical history. She denies prior illnesses. She had a prior occupational injury in ____ when she worked for ________ when she jumped down from a machine and slipped on a popsicle on the floor, hitting her back on the stairs. She tells me she injured her neck and back during that accident and bruised her lung. She treated with Dr. _____ , a Chiropractor at that time. She says she was off work for 4 and a half years as a result of that accident.

She has a prior motor vehicle accident in ______ when she was the back seat passenger in a car, the vehicle was rear ended by another driver. She was taken to a hospital in _____ She treated with a Chiropractor, Dr. in _______ from _____ She has had no follow up care since that time regarding that accident.

Figure 7-34 Independent chiropractic exam (ICE) II.

Impression: Cervical, thoracic and lumbar sprain strain syndrome, spasm, joint dysfunction, possible fractured ribs. Possible manubrium fracture.

Figure 7-35 Independent chiropractic exam (ICE) III regarding rib fracture.

Need for diagnostic testing: I cannot rule out the need for further diagnostic testing at this time, especially in the form of plain film x-rays of the thoracic spine, ribs and manubrium to rule out (o rather verify, rule in) healing status of fractures. I cannot rule out the need for additional advance diagnostic testing at this time, but cannot recommend it at this time either.

Disability: The claimant was not employed at the time of the accident, thus disability status is not an issue. However, I do not identify any element of disability on today's examination that would stop her from seeking employment.

Figure 7-36 Independent chiropractic exam (ICE) IV regarding disability.

Causal Relationship: With the exclusion of suspected fractures, I find causal relationship to the accident of ______ and her current symptoms. While it is very possible (and probable) that she may have sustained fractures during this accident, all plain film testing does not verify this. Bone scan is sensitive to fracture, but not specific to fracture, thus, until I can see proof otherwise, I cannot find causal relationship to fracture and the ______ accident.

Figure 7-37 *Independent chiropractic exam (ICE) V regarding causality.*

Cervical MRI imaging (Figures 7-38 and 7-39) was performed to address the patient's complaint of numbness and tingling in the right arm. The radiologist's impression was of small to moderate distortion of all discs, C2 through C7.

IMPRESSION: MILD DIFFUSE POSTERIOR BULGE OF THE ANNULUS FIBROSUS AT C2-3.

SMALL TO MODERATE SIZE CENTRAL DISC PROTRUSION AT C3-4.

VERY SMALL C4-5 CENTRAL DISC PROTRUSION WITHOUT MASS EFFECT UPON THE CERVICAL CORD OR NEURAL FORAMINAL COMPROMISE.

MODERATE SEVERITY CIRCUMFERENTIAL BULGE OF THE ANNULUS FIBROSUS AT C5-6.

MODERATE SEVERITY RIGHT PARACENTRAL FOCAL DISC PROTRUSION AT C6-7.

Thank you for your kind referral.

Figure 7-38 *Impression of the cervical MRI.*

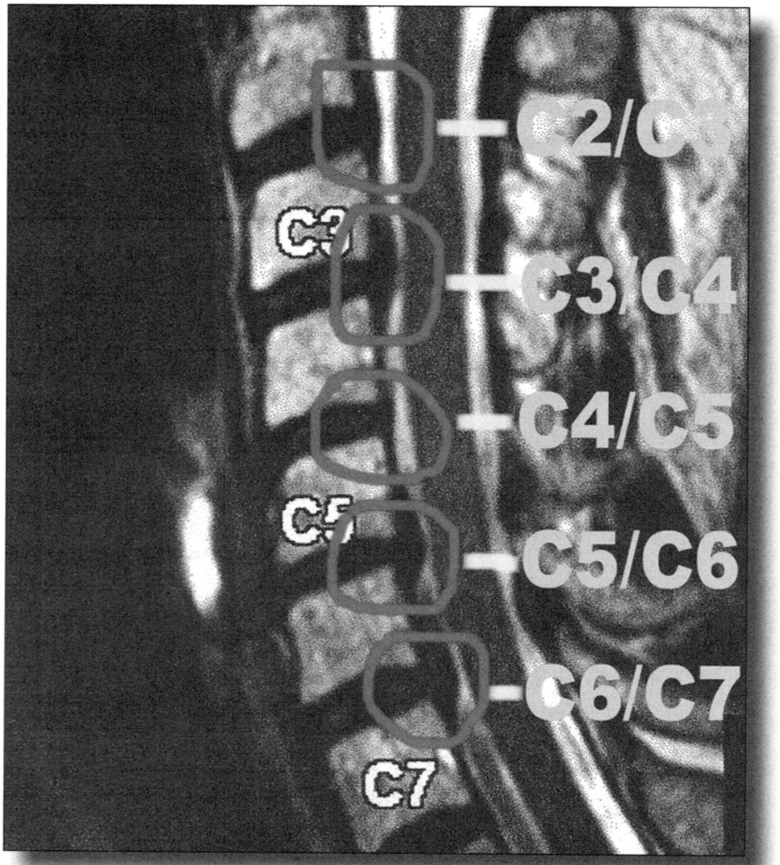

Figure 7-39 *Cervical MRI image.*

The last document is a time line (Figure 7-40). In many instances, such a document can be a useful way of organizing the various events and/or treatments and possibly may provide some insight into causality and the contribution of pre-existing conditions.

Summary

In summary, there are well-documented anomalies of the cervical and lumbar regions of the spine, and these would be expected to foster arthritis and change tolerance to trauma. There also is a history of previous and significant related injuries. The records present a wide range of histories, and one might speculate the extent to which the various clinicians were influenced in their impressions by the histories they received. Likewise, there was a rather unusual confrontation with the ER nurse, and one could speculate regarding the reliability of

Time Line (Events & Treatment)

0	Congenital	Partial fusion - C2/3, only 4 lumbar vert.
1	1/1/1990	Occupational injury, off work 4-1/2 yrs.
2	1/1/2000	mva, rear-ended, treated for 6 mos.
3	1/1/2002	Subject MVA, extract & take to ER, XR's
4	1/9/2002	DC1 OV, C/T/L XR's
5	1/13/2002	Bone Scan, Chest XR's
6	2/17/2002	DC2 OV
7	4/9/2002	IME
8	5/27/2002	Neck MRI

***Figure** 7-40 Time line.*

that history. Apparently, there were two separate impacts—the first one to the driver door, and the second one a rear impact. There was a fracture of the manubrium bone of the sternum and most likely bilateral rib injury, more pronounced on the left than on the right. This would be consistent with chest compression and possibly chest concentrated loading. The rib/sternum injury pattern as identified on the body scan does not seem consistent with the "upper left and lower right" pattern one would expect from the driver's shoulder belt and more likely would be consistent with loading from other components of the vehicle interior, such as the side door, airbag, or steering wheel. The widespread and congenital spinal structural anomalies make the spine susceptible to degenerative changes and injury. Furthermore, the extensive bulging cervical discs, without any indication of bony or ligamentous injury, probably are not due to the most recent MVA. In the absence of specific before and after radiologic imaging, I would be reluctant to ascribe any of the observed spinal conditions to this MVA. Furthermore, in my opinion, the medical history is too cloudy to assess the contribution of the most recent MVA with regard to sprain/strain injury to the cervical, thoracic, and lumbar regions. The bruising and tenderness of the left hip region and the rib and sternal fractures are attributable to this MVA.

Chapter 8

א

Advanced Case Study II: High-Speed Rear Impact

Introduction

This advanced case study provides a number of records (Figures 8-1 through 8-18). The goal is to review them and to integrate them to form working hypotheses regarding identifying any significant injuries and their cause(s). The records are arranged in groups, namely, emergency room (ER) (Figure 8-1), emergency medical services (EMS) (Figure 8-2), radiology report (Figure 8-3), police report (Figure 8-4), toxicology report (Figures 8-5 and 8-6), autopsy (Figures 8-7 through 8-17), and death certificate (Figure 8-18).

Note: A discussion follows Figure 8-18. After that discussion, the 18 figures are repeated, with some callouts and/or marginalia added as well as some additional figures. For those readers who wish to try the analysis on their own first, it is recommended that they not look past Figure 8-18 until they have tried the exercise. For those who prefer another "guided tour," it is recommended they go directly to the Discussion section and the second set of figures.

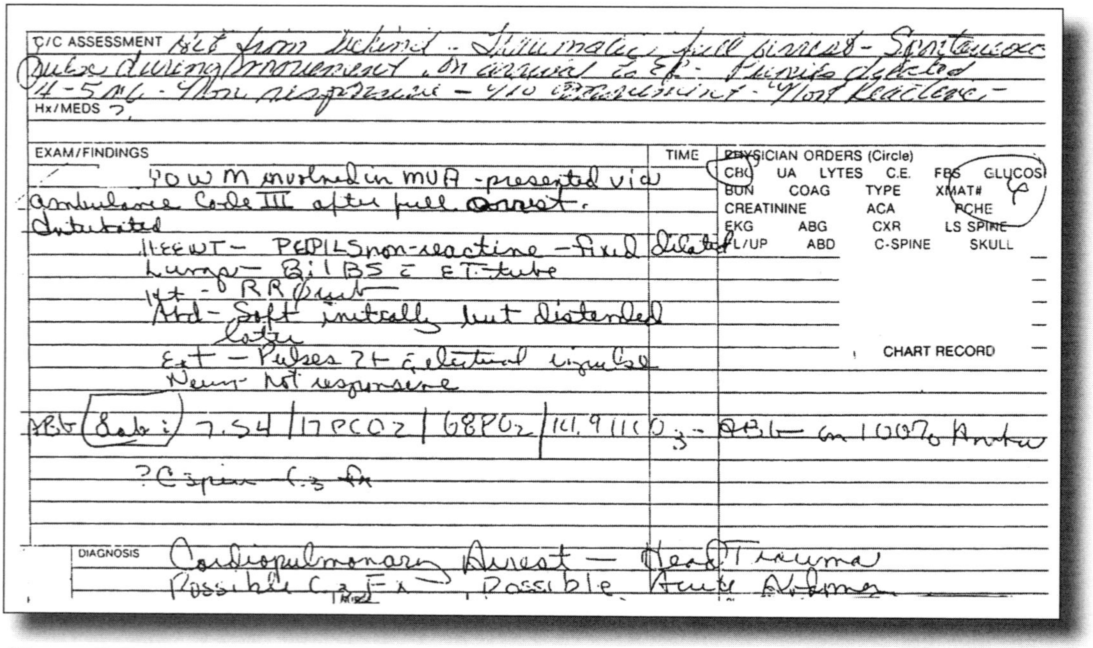
C/C ASSESSMENT Hit from behind - Traumatic full arrest - Spontaneous pulse during movement. On arrival to ER - Pupils dilated 4-5 mm - Non responsive - 4/10 movement - Not reactive -
Hx/MEDS ?
EXAM/FINDINGS
40 WM involved in MVA - presented via ambulance Code III after full arrest. Intubated
HEENT - PUPILS non-reactive - fixed dilated
Lungs - Bil BS c ET tube
Ht - RR
Abd - Soft initially but distended later
Ext - Pulses 2+ c electrical impulse
Neuro - Not responsive
ABG (Lab:) 7.54 / 17 PCO2 / 68 PO2 / 14.9 HCO3 - ABG on 100% Ambu
? C spine C3 fx
TIME
PHYSICIAN ORDERS (Circle)
CBC UA LYTES C.E. FBS GLUCOSE
BUN COAG TYPE XMAT# PCHE
CREATININE ACA
EKG ABG CXR LS SPINE
FL/UP ABD C-SPINE SKULL
CHART RECORD
DIAGNOSIS Cardiopulmonary Arrest - Head Trauma
Possible C3 Fx, Possible Acute Abdomen

Figure 8-1 *Emergency room (ER) history I.*

I asked [redacted] if he recalled if Dr #1 was wearing his seat or shoulder belts. He stated that he does not recall that Dr #1 was wearing either. I asked him if he could recall removing any belts restraints or cutting any belt restraints in order to remove Dr #1 and he stated that he did not remember any of this and reiterated that he did not recall if Dr #1 was wearing any seat or shoulder restraints and said "I don't believe he was wearing any".

Figure 8-2 *Emergency medical services (EMS) report.*

RADIOLOGY REPORT

PCXR,Port c-spi re 1 v

EXAMINATION ______________________________

CHEST: Mobile exam at the bedside; AP supine projection.

An endotracheal tube tip is located in the distal third of the trachea. The remaining cardiac and pulmonary findings are within normal limits.

CERVICAL SPINE: A single lateral projection.

C1 through C5 are visualized, with C6 and 7 unvisualized for adequate evaluation. There is no gross evidence to suggest fracture or osseous pathology.

Figure 8-3 *Radiology report.*

As Dr #1 negotiated a sliaht curve his sedan began to slide across the center line of this four-lane roadway. The vehicle then reversed ends and slid, rear first, into the westbound lanes, colliding with the van. The left rear of vehicle #1 struck the front center of the van. After impact, the #1vehicle came to rest in a shallow ditch on the north side facing east. Evidently, Vehicle #1 made a full 360 degree spin. The van remained in the westbound lanes, facing slightly northwest.

Figure 8-4 *Police report.*

TOXICOLOGY TEST RESULTS

OFFICE OF CHIEF MEDICAL EXAMINER
TOXICOLOGY LABORATORY SERVICE

M.D., DABFP
CHIEF MEDICAL EXAMINER

CHIEF TOXICOLOGIST

NAME:
M.E. CASE NUMBER:

PRIORITY:
M.E. TOX NUMBER:

BLOOD	ETHANOL	NEG
VITREOUS	ETHANOL	NEG
HOSP BLOOD	ETHANOL	NEG
HOSP URINE	ETHANOL	NEG

EVIDENCE TRANSMITTAL DATA

DATE: TIME:
RELEASED BY:
RECEIVED BY:

APPROVED BY:
APPROVED DATE:

Figure 8-5 *Toxicology report.*

BLOOD	ETHANOL	NEG
VITREOUS	ETHANOL	NEG
HOSP BLOOD	ETHANOL	NEG
HOSP URINE	ETHANOL	NEG

Figure 8-6 *Toxicology report—close-up.*

OFFICE OF CHIEF MEDICAL EXAMINER

AUTOPSY

Case No:

Approximate Age: Weight: lbs.

Height: .nches Sex:

Identified by: See medical investigator report.

Stenographer:

I hereby certify that I, ____________, M. D., have performed an autopsy on the body of ____________ at the County Morgue on the ____________ and said autopsy revealed:

FINDINGS:

I. Blunt force trauma from motor vehicle - motor vehicle collision (driver), with:
 A. Diffuse cerebral injury, with:
  1. Acute cerebral edema.
  2. Brainstem herniation.
  3. Midline pontine hemorrhages.
  4. Small subdural hematoma, inferior.
 B. Mild to moderate atlanto-occipital subluxation with surrounding soft tissue hemorrhages.
 C. Anterior body fracture to cervical vertebra C7, with surrounding soft tissue hemorrhages.
 D. Contusion to lungs, bilateral lower lobes.
 E. Large peri-psoas hematoma, bilateral.
 F. Skeletal fractures:
  1. Ribs #3 through 8, right anterolateral.
  2. Ribs #7 and 8, left posterolateral.
  3. Atlanto-occipital subluxation, moderate.
  4. Fracture, anterior body of C7 cervical vertebra.
 G. Multiple cutaneous contusions and abrasions.

II. Moderate anthracosis, lungs, bilateral.

III. Status post organ harvesting:
 A. Heart.
 B. Liver and gallbladder.
 C. Kidneys, bilateral.
 D. Spleen.
 E. Adrenal glands, with residual right fragment recovered.

CAUSE OF DEATH: CRANIOCEREBRAL TRAUMA DUE TO BLUNT FORCE TRAUMA FROM MOTOR VEHICLE - MOTOR VEHICLE COLLISION (DRIVER)

Figure 8-7 *Autopsy report.*

I. Blunt force trauma from motor vehicle - motor vehicle collision (driver), with:
    A. Diffuse cerebral injury, with:
        1. Acute cerebral edema.
        2. Brainstem herniation.
        3. Midline pontine hemorrhages.
        4. Small subdural hematoma, inferior.
    B. Mild to moderate atlanto-occipital subluxation with surrounding soft tissue hemorrhages.
    C. Anterior body fracture to cervical vertebra C7, with surrounding soft tissue hemorrhages.

Figure 8-8 *Autopsy—head and neck.*

SKELETAL INJURIES: The skull varies between 4 and 12 mm. in thickness and shows no evidence of fracturing.

There is a mild pattern of atlanto-occipital subluxation with hemorrhage extending into the soft tissues of the neck surrounding cervical vertebrae C1 through C3. There is fracturing between the body of C7 and the intervertebral disc to T1 with extensive hemorrhages in the lower neck soft tissues surrounding this. There is a pattern of fracturing of ribs #3 through 8, right anterolateral and ribs #7 and 8, left posterior. There is a pattern of extensive contusion to the intercostal spaces 4 through 7 in the left posterolateral chest, that is roughly correspondent with the cutaneous contusion that has been described previously.

Figure 8-9 *Autopsy—fractures.*

Skeletal fractures:
1. Ribs #3 through 8, right anterolateral.
2. Ribs #7 and 8, left posterolateral.
3. Atlanto-occipital subluxation, moderate.
4. Fracture, anterior body of C7 cervical vertebra.

Figure 8-10 *Autopsy—skeletal fractures.*

The scalp reveals a full amount of straight brown hair 3 1/2 inches in maximum length, receded 2 inches from the anterior hairline at the temple regions bilaterally. The face is clean shaven. There is an average amount of darkly pigmented body hair present. The scalp shows no external injury pattern. The eyes are closed with clear corneae, moderately congested conjunctival and scleral membranes, and brown irides that subtend equal pupil diameters at 6 mm. Arcus senilis and cataract are not present. The orbits are normally formed. The whole eyes have been removed. The nose is unremarkable.

Figure 8-11 *Autopsy—scalp.*

DESCRIPTION OF BLUNT FORCE INJURIES:

CUTANEOUS INJURIES: In the base of the chin, there is over toward the left side, a 2 x 3/4 inch area of faint purple contusion.

Figure 8-12 *Autopsy— blunt force injuries.*

Within the left shoulder there is a 7 x 3 inch prominent area of dark purple fresh contusion. At the lower outer quadrant of the pectoral region there is a 1 1/2 inch area of dark purple fresh contusion. At the anterior surface of the upper right arm there is a 1 1/2 x 1 inch area of dark purple contusion. The dorsal surface of the left wrist reveals a 1 1/4 inch area of very superficial gliding abrasion produced by proximal forces. The anterior tibial surface on the right reveals a 1 x 1/4 inch wide contact abrasion with a similar contact abrasion 3/4 x 1/4 inch noted in the left anterior tibial surface.

The left posterior shoulder reveals a 2 1/2 x 1 inch area of dark purple fresh contusion.

Figure 8-13 *Autopsy—shoulder.*

INTERNAL INJURIES: There is a mild pattern of inferior surface subdural hematoma approximating 50 cc., with a pattern of early clotting and no significant adherence to the membranes. There is a very prominent pattern of cerebral edema with flattening of the gyri and crowding of the sulci and a pattern of deep uncal grooving and flattening and early necrosis to the cerebellar tonsils. No cingulate deviation is observed. The inferior surface of the frontal lobes bilaterally reveal mild patterns of gyral tip contusions. The cerebrum is moderately soft. There is edema and softness and swelling to the cerebellum with a pattern of hemorrhage into the dentate nucleus on the left. The velum of the pontine brainstem reveals a pattern of hemorrhages with the brainstem tissue being soft and edematous. The ventricles are compressed with minimal cerebrospinal fluid remaining.

Lower lobes to both lungs reveal a pattern of posterior contusions. There is an internal pattern of bilateral peri-psoas hematoma occupying the region between T9 and L3, with an overall pattern of 35 cm. in length x 14 cm. in width. The cutaneous contusion observed in the left shoulder extends into the deep underlying muscle and soft tissue at the left shoulder as well.

Figure 8-14 *Autopsy—internal injuries.*

Lower lobes to both lungs reveal a pattern of posterior contusions. There is an internal pattern of bilateral peri-psoas hematoma occupying the region between T9 and L3, with an overall pattern of 35 cm. in length x 14 cm. in width. The cutaneous contusion observed in the left shoulder extends into the deep underlying muscle and soft tissue at the left shoulder as well.

INTERNAL EXAMINATION: A standard Y-shaped thoraco-abdominal incision is utilized as an extension of the already present midline incision. A standard transparietal scalp incision is also utilized.

SEROUS CAVITIES: There is thin watery blood tinged fluid present within the serous cavities to a moderate degree. Marked internal visceral pallor is present, in the status post organ harvesting state.

Figure 8-15 *Autopsy—lungs.*

CENTRAL NERVOUS SYSTEM: The brain weighs 1477 grams. There is no evidence of subgaleal scalp injury. Subdural hematoma has been previously described. The subarachnoid veins reveal a pattern of prominent congestion and there is a small amount of blood tingeing within the compressed sulci. Prominent edema has been described. The arterial system is intact and without atherosclerotic change. Venous sinuses contain clotted blood. Architectural outlines are unremarkable and serial sections through the cerebrum, cerebellum and brainstem. The early necrotic changes have been described. Cut sections also reveal a pattern of small vessel prominent congestion throughout. Ventricular compression has been described. The spinal cord appears intact and without evidence of significant injury.

Figure 8-16 *Autopsy—central nervous system (CNS).*

MANNER OF DEATH: ACCIDENTAL

M. D.
Deputy Medical Examiner

Figure 8-17 *Autopsy—manner of death.*

CAUSE OF DEATH

28. PART I. Enter the diseases, injuries, or complications that caused the death. Do not enter the mode of dying, such as cardiac or respiratory arrest, shock, or heart failure. List only one cause on each line.

Approximate Interval Between Onset and Death

IMMEDIATE CAUSE (Final disease or condition resulting in death) → a. CEREBRAL TRAUMA

DUE TO (OR AS A LIKELY CONSEQUENCE OF):

b. MOTOR VEHICLE-MOTOR VEHICLE COLLISION (DRIVER) — Hours

Sequentially list conditions, if any, leading to immediate cause. Enter UNDERLYING CAUSE (Disease or injury that initiated events resulting in death) LAST

DUE TO (OR AS A LIKELY CONSEQUENCE OF):

c.

DUE TO (OR AS A LIKELY CONSEQUENCE OF):

d.

PART II. Other significant conditions contributing to death but not resulting in the underlying cause given in Part I.

30a. WAS AN AUTOPSY PERFORMED? ☒YES ☐NO

30b. WERE AUTOPSY FINDINGS AVAILABLE PRIOR TO COMPLETION OF CAUSE OF DEATH? ☒YES ☐NO

29a. Was decedent pregnant at time of death? ☐YES ☒NO ☐UNKNOWN

29b. Was decedent pregnant during the last 12 months? ☐YES ☒NO ☐UNKNOWN

Figure 8-18 *Death certificate.*

Note: Immediately following this note, there is a discussion of the case, followed by a repeat of the 18 figures (i.e., Figure 8-19 is a repeat of Figure 8-1, Figure 8-20 is a repeat of Figure 8-2, and so on). However, this second presentation of the "original" figures (Figures 8-19 through 8-36) includes callouts and/or marginalia, as well as some additional figures (Figure 8-37 and 8-38) providing background material. For those readers who wish to try the analysis on their own first, it is recommended that they not read the Discussion until they have tried the exercise. For those who prefer another "guided tour," it is recommended they go directly to the Discussion section and the second set of figures.

Discussion

This emergency room (ER) record (Figure 8-19) provides some basic details regarding the history of the event and the patient. Under "CC [Chief Complaint]/Assessment," the record indicates the vehicle was "hit from behind." In the "Exam/Findings" section, the record indicates that this patient was transported from the accident scene via ambulance, "Code 3" (reserved for the most severe injuries) after full arrest, that is, the patient had stopped breathing, and in the "HEENT" (head, eyes, ears, nose, and throat) subsection, the record indicates "PUPILS non-reactive—fixed dilated" (indicative of severely compromised neurologic function). This record also raises the question of a fracture of C3. Although this fracture, per se, would be a relatively minor injury, it may prove informative regarding the injury mechanism.

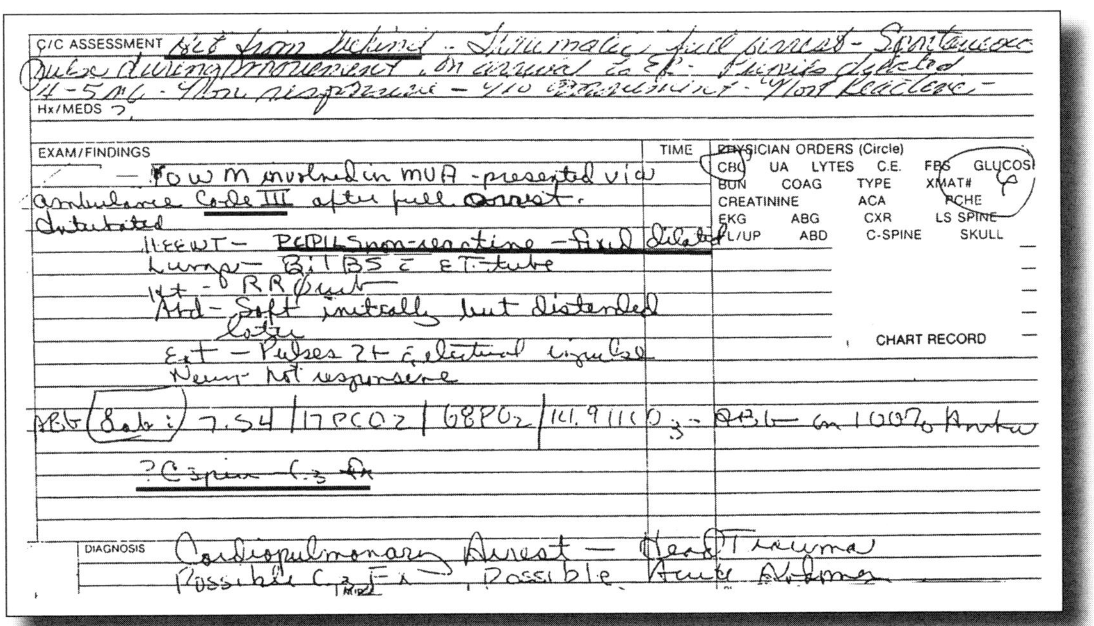

Figure 8-19 Emergency room (ER) history I.

The next record (Figure 8-20) is from an interview with one of the EMS personnel. With regard to addressing if the patient, the driver of the vehicle, was belted, the EMS person did not recall if the driver was seat belted nor did he recall having to cut him out of the seat belt. Upon further reflection, the EMS person stated that he didn't "believe he [the patient] was wearing any [seat belt]."

I asked [redacted] if he recalled if Dr #1 was wearing his seat or shoulder belts. He stated that he does not recall that Dr #1 was wearing either. I asked him if he could recall removing any belts restraints or cutting any belt restraints in order to remove Dr #1 and he stated that he did not remember any of this and reiterated that he did not recall if Dr #1 was wearing any seat or shoulder restraints and said "I don't believe he was wearing any".

Figure 8-20 *Emergency medical services (EMS) report.*

The next record (Figure 8-21), a radiology report of a cervical x-ray taken at bedside, does not report any fracture. However, the report notes that only C1 through C5 are adequately visualized and that the x-ray is not adequate to evaluate C6 and C7.

RADIOLOGY REPORT

PCXR,Port c-spine 1 v

EXAMINATION ______

CHEST: Mobile exam at the bedside; AP supine projection.

An endotracheal tube tip is located in the distal third of the trachea. The remaining cardiac and pulmonary findings are within normal limits.

CERVICAL SPINE: A single lateral projection.

C1 through C5 are visualized, with C6 and 7 unvisualized for adequate evaluation. There is no gross evidence to suggest fracture or osseous pathology.

Figure 8-21 *Radiology report.*

This police accident report (PAR) (Figure 8-22) provides a basic description of the accident scenario. This vehicle was involved in a high-speed rear impact, but one that resulted from the subject vehicle spinning and sliding into the lane for opposing traffic and effectively sliding rear-end first into an oncoming van. Thus, rather than the front of the van impacting the rear of the patient's vehicle, the rear of the patient's vehicle impacted the front of the van. This impact scenario is in marked contrast to that conveyed by the wording in the ER record as shown as Figure 8-19 ("hit from behind").

As Dr #1 negotiated a slight curve his sedan began to slide across the center line of this four-lane roadway. The vehicle then reversed ends and slid, rear first, into the westbound lanes, colliding with the van. The left rear of vehicle #1 struck the front center of the van. After impact, the #1vehicle came to rest in a shallow ditch on the north side facing east. Evidently, Vehicle #1 made a full 360 degree spin. The van remained in the westbound lanes, facing slightly northwest.

Figure 8-22 *Police report.*

Figures 8-23 and 8-24 provide a summary of toxicology test results. Samples of blood, vitreous (from the eye), and urine were tested and gave consistent results, namely, they were negative for alcohol. Note that this document also provides a record of evidence custody (Figure 8-23).

Figures 8-25 through 8-35 are from the autopsy report.

The first two pages (Figures 8-25 and 8-26) under "Findings" provide a description of head and neck injury. The blunt force trauma listed in Item I (Figure 8-26) includes cerebral and spine injury. Head injury includes cerebral edema, brainstem herniation, hemorrhages, and sub-dural hematoma. (Those wishing more background discussion regarding these types of injuries are referred to Pike, 1990.) The cerebral edema or accumulation of fluid becomes especially problematic when the brain swelling deforms the brain substance and/or reaches an extent that the skull no longer provides sufficient space. When there is bleeding within the skull, the blood may accumulate on either side of the dura mater—the tough, outermost protective membrane surrounding the brain. When blood collects beneath the dura, that is, on the "brain side" rather than on the "skull side" of the dura, the condition is labeled as a

TOXICOLOGY TEST RESULTS

OFFICE OF CHIEF MEDICAL EXAMINER
TOXICOLOGY LABORATORY SERVICE

M.D. CABF
CHIEF MEDICAL EXAMINER

CHIEF TOXICOLOGIST

NAME:
M.E. CASE NUMBER:

PRIORITY:
M.E. TOX NUMBER:

BLOOD	ETHANOL	NEG
VITREOUS	ETHANOL	NEG
HOSP BLOOD	ETHANOL	NEG
HOSP URINE	ETHANOL	NEG

EVIDENCE TRANSMITTAL DATA

DATE: TIME:
RELEASED BY:
RECEIVED BY:

APPROVED BY:
APPROVED DA

Figure 8-23 *Toxicology report.*

BLOOD	ETHANOL	NEG
VITREOUS	ETHANOL	NEG
HOSP BLOOD	ETHANOL	NEG
HOSP URINE	ETHANOL	NEG

Figure 8-24 *Toxicology report—close-up.*

sub-dural hematoma. (Blood collection on the "skull side" is termed epidural hematoma.) If the blood continues to accumulate, the mass starts to push against the brain, thereby producing local deformation, tending to compress and displace brain tissue. Note that Figure 8-25, under Item III, notes harvested organs, and at the bottom of this page, the cause of death is given as "Craniocerebral trauma due to blunt force trauma from motor vehicle–motor vehicle collision (driver)."

OFFICE OF CHIEF MEDICAL EXAMINER

AUTOPSY

Case No:

Approximate Age: Weight: lbs.

Height: inches Sex:

Identified by: See medical investigator report.

Stenographer:

I hereby certify that I, ______________, M. D., have performed an autopsy on the body of __________ at the County Morgue on the ______________ and said autopsy revealed:

FINDINGS:

I. Blunt force trauma from motor vehicle - motor vehicle collision (driver), with:
   A. Diffuse cerebral injury, with:
      1. Acute cerebral edema.
      2. Brainstem herniation.
      3. Midline pontine hemorrhages.
      4. Small subdural hematoma, inferior.
   B. Mild to moderate atlanto-occipital subluxation with surrounding soft tissue hemorrhages.
   C. Anterior body fracture to cervical vertebra C7, with surrounding soft tissue hemorrhages.
   D. Contusion to lungs, bilateral lower lobes.
   E. Large peri-psoas hematoma, bilateral.
   F. Skeletal fractures:
      1. Ribs #3 through 8, right anterolateral.
      2. Ribs #7 and 8, left posterolateral.
      3. Atlanto-occipital subluxation, moderate.
      4. Fracture, anterior body of C7 cervical vertebra.
   G. Multiple cutaneous contusions and abrasions.

II. Moderate anthracosis, lungs, bilateral.

III. Status post organ harvesting:
   A. Heart.
   B. Liver and gallbladder.
   C. Kidneys, bilateral.
   D. Spleen.
   E. Adrenal glands, with residual right fragment recovered.

CAUSE OF DEATH: CRANIOCEREBRAL TRAUMA DUE TO BLUNT FORCE TRAUMA FROM MOTOR VEHICLE - MOTOR VEHICLE COLLISION (DRIVER)

Figure 8-25 *Autopsy report.*

```
I.    Blunt force trauma from motor vehicle - motor vehicle
      collision (driver), with:
      A.    Diffuse cerebral injury, with:
            1.    Acute cerebral edema.
            2.    Brainstem herniation.
            3.    Midline pontine hemorrhages.
            4.    Small subdural hematoma, inferior.
      B.    Mild to moderate atlanto-occipital subluxation with
            surrounding soft tissue hemorrhages.
      C.    Anterior body fracture to cervical vertebra C7, with
            surrounding soft tissue hemorrhages.
```

Figure 8-26 *Autopsy—head and neck.*

Figures 8-26 and 8-27 describe an atlanto-occipital subluxation, that is, a partial separation of the skull from the top of the spinal column (C1), which is indicative of a force with a significant tensional component, and a fracture of the anterior of C7, which would be indicative of a neck flexion injury mechanism. (Note that the neck x-ray [Figure 8-21] was inadequate to assess C7 and indeed did not detect the C7 fracture.)

```
SKELETAL INJURIES:  The skull varies between 4 and 12 mm. in
thickness and shows no evidence of fracturing.

There is a mild pattern of atlanto-occipital subluxation  with
hemorrhage extending into the soft tissues of the neck surrounding
cervical vertebrae C1 through C3.  There is fracturing between the
body of C7 and the intervertebral disc to T1 with extensive
hemorrhages in the lower neck soft tissues surrounding this.  There
is a pattern of fracturing of ribs #3 through 8, right
anterolateral and ribs #7 and 8, left posterior.  There is a
pattern of extensive contusion to the intercostal spaces 4 through
7 in the left posterolateral chest, that is roughly correspondent
with the cutaneous contusion that has been described previously.
```

Figure 8-27 *Autopsy—fractures.*

Figures 8-27 and 8-28 identify fractures of the upper and middle ribs on both the left and right sides. The fracture pattern does not conform to a normally positioned shoulder belt and, depending on related information, may be one indicator of the seat belt not being used.

```
Skeletal fractures:
1.    Ribs #3 through 8, right anterolateral.
2.    Ribs #7 and 8, left posterolateral.
3.    Atlanto-occipital subluxation, moderate.
4.    Fracture, anterior body of C7 cervical vertebra.
```

Figure 8-28 *Autopsy—skeletal fractures.*

The next section of the autopsy (Figure 8-29) includes the scalp. It notes that the scalp "shows no external injury pattern," thereby starting to support the notion that there was not a significant head impact. Similarly, the notation that the "nose is unremarkable" is another indicator (i.e., because the nose was not fractured) that there was not significant mid-face impact trauma. There also was mention that the eyes had been removed, which, of course, would be reflective of an anatomical donor program.

```
The scalp reveals a full amount of straight brown hair 3 1/2 inches
in maximum length, receded 2 inches from the anterior hairline at
the temple regions bilaterally.  The face is clean shaven.  There
is an average amount of darkly pigmented body hair present.  The
scalp shows no external injury pattern.  The eyes are closed with
clear corneae, moderately congested conjunctival and scleral
membranes, and brown irides that subtend equal pupil diameters at
6 mm.  Arcus senilis and cataract are not present.  The orbits are
normally formed.  The whole eyes have been removed.  The nose is
unremarkable.
```

Figure 8-29 *Autopsy—scalp.*

The next autopsy section (Figure 8-30) provides a good example of an injury that may seem extremely minor clinically but provides some important information regarding the possible injury mechanism. The contusion under the chin would seem to support the notion of an "upper cut" extension-type neck injury and may appear to be in contradiction to the flexion-type vertebral fracture. However, a more inclusive and alternative interpretation would be that the chin was bruised as a result of chin contact (with the upper chest) caused by the neck flexion.

DESCRIPTION OF BLUNT FORCE INJURIES:

CUTANEOUS INJURIES: In the base of the chin, there is over toward the left side, a 2 x 3/4 inch area of faint purple contusion.

Figure 8-30 *Autopsy—blunt force injuries.*

The next section of the autopsy (Figure 8-31) discusses the left shoulder, right arm, and shins. It is noted that both shins demonstrated contact abrasions. Depending on other crash details—including the impact speed and direction, seating position, driver height, and vehicle interior contact marks—the shin abrasions may be indicative of forward motion of the driver (knee/shin contact with the lower instrument panel) and may be relevant to the issue of seat belt usage.

Within the left shoulder there is a 7 x 3 inch prominent area of dark purple fresh contusion. At the lower outer quadrant of the pectoral region there is a 1 1/2 inch area of dark purple fresh contusion. At the anterior surface of the upper right arm there is a 1 1/2 x 1 inch area of dark purple contusion. The dorsal surface of the left wrist reveals a 1 1/4 inch area of very superficial gliding abrasion produced by proximal forces. The anterior tibial surface on the right reveals a 1 x 1/4 inch wide contact abrasion with a similar contact abrasion 3/4 x 1/4 inch noted in the left anterior tibial surface.

The left posterior shoulder reveals a 2 1/2 x 1 inch area of dark purple fresh contusion.

Figure 8-31 *Autopsy—shoulder.*

The next section of the autopsy (Figure 8-32) is titled "Internal Injuries." This record notes extensive bruising in the mid-back and a deep bruising in the left shoulder region. The record also provides more detail of the brain injuries mentioned earlier (Figure 8-26). The observation that the "ventricles are compressed" is one more indication that the combination of brain swelling and intracranial hematoma is producing sufficient distortion of neural tissue to reduce the space in the ventricles. The autopsy also notes that "the inferior surface of the frontal lobes bilaterally reveal mild patterns of gyral tip contusions."

```
INTERNAL INJURIES:  There is a mild pattern of inferior surface
subdural hematoma approximating 50 cc., with a pattern of early
clotting and no significant adherence to the membranes.  There is
a very prominent pattern of cerebral edema with flattening of the
gyri and crowding of the sulci and a pattern of deep uncal grooving
and flattening and early necrosis to the cerebellar tonsils.  No
cingulate deviation is observed.  The inferior surface of the
frontal lobes bilaterally reveal mild patterns of gyral tip
contusions.  The cerebrum is moderately soft.  There is edema and
softness and swelling to the cerebellum with a pattern of
hemorrhage into the dentate nucleus on the left.  The velum of the
pontine brainstem reveals a pattern of hemorrhages with the
brainstem tissue being soft and edematous.  The ventricles are
compressed with minimal cerebrospinal fluid remaining.

Lower lobes to both lungs reveal a pattern of posterior contusions.
There is an internal pattern of bilateral peri-psoas hematoma
occupying the region between T9 and L3, with an overall pattern of
35 cm. in length x 14 cm. in width.  The cutaneous contusion
observed in the left shoulder extends into the deep underlying
muscle and soft tissue at the left shoulder as well.
```

Figure 8-32 *Autopsy— internal injuries.*

Note that in Figure 8-33, some of the findings are attributed to organ harvesting rather than traumatic injury.

```
Lower lobes to both lungs reveal a pattern of posterior contusions.
There is an internal pattern of bilateral peri-psoas hematoma
occupying the region between T9 and L3, with an overall pattern of
35 cm. in length x 14 cm. in width.  The cutaneous contusion
observed in the left shoulder extends into the deep underlying
muscle and soft tissue at the left shoulder as well.

INTERNAL EXAMINATION:  A standard Y-shaped thoraco-abdominal
incision is utilized as an extension of the already present midline
incision.  A standard transparietal scalp incision is also
utilized.

SEROUS CAVITIES:  There is thin watery blood tinged fluid present
within the serous cavities to a moderate degree.  Marked internal
visceral pallor is present, in the status post organ harvesting
state.
```

Figure 8-33 *Autopsy—lungs.*

Figure 8-34 focuses on the central nervous system (CNS). It indicates that there is no evidence of subgaleal scalp injury, that is, no injury to the middle layer of the scalp. This is one indicator that there was not significant head impact. This record also notes the sub-dural hematoma. Because this type of hematoma typically is associated with deceleration rather than head contact per se, it may be taken as further support of a brain injury scenario that does not include a violent head impact. This record likewise notes that the spinal cord appears intact without evidence of significant injury.

CENTRAL NERVOUS SYSTEM: The brain weighs 1477 grams. There is no evidence of subgaleal scalp injury. Subdural hematoma has been previously described. The subarachnoid veins reveal a pattern of prominent congestion and there is a small amount of blood tingeing within the compressed sulci. Prominent edema has been described. The arterial system is intact and without atherosclerotic change. Venous sinuses contain clotted blood. Architectural outlines are unremarkable and serial sections through the cerebrum, cerebellum and brainstem. The early necrotic changes have been described. Cut sections also reveal a pattern of small vessel prominent congestion throughout. Ventricular compression has been described. The spinal cord appears intact and without evidence of significant injury.

Figure 8-34 *Autopsy—central nervous system (CNS).*

The next record (Figure 8-35) states that the manner of death was "accidental." In many instances, this single word description of the manner of death may be the most significant autopsy assessment—implicit in "accidental" is the determination that it was not homicide, suicide, and so forth.

MANNER OF DEATH: ACCIDENTAL

M. D.
Deputy Medical Examiner

Figure 8-35 *Autopsy—manner of death.*

The next record (Figure 8-36) is from the death certificate. It specifies cerebral trauma as the immediate cause (of death) and then indicates that this was due to a "motor vehicle–motor vehicle collision," that the deceased was the driver of one of the vehicles, and that death occurred within hours of the collision. There also is the indication that an autopsy was performed and that the autopsy findings were available prior to completion of the cause of death on the death certificate.

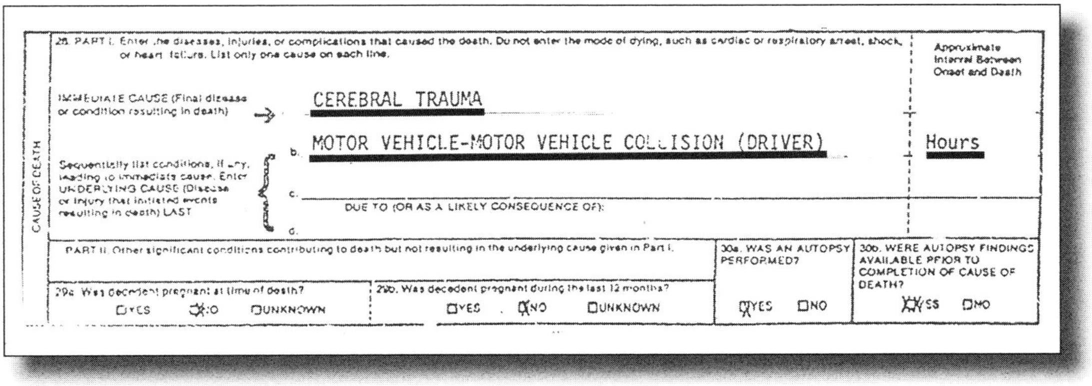
CAUSE OF DEATH

28. PART I. Enter the diseases, injuries, or complications that caused the death. Do not enter the mode of dying, such as cardiac or respiratory arrest, shock, or heart failure. List only one cause on each line.

Approximate Interval Between Onset and Death

IMMEDIATE CAUSE (Final disease or condition resulting in death) → CEREBRAL TRAUMA

Sequentially list conditions, if any, leading to immediate cause. Enter UNDERLYING CAUSE (Disease or injury that initiated events resulting in death) LAST

b. MOTOR VEHICLE-MOTOR VEHICLE COLLISION (DRIVER) — Hours

c.

DUE TO (OR AS A LIKELY CONSEQUENCE OF):

d.

PART II. Other significant conditions contributing to death but not resulting in the underlying cause given in Part I.

30a. WAS AN AUTOPSY PERFORMED? ☒YES ☐NO

30b. WERE AUTOPSY FINDINGS AVAILABLE PRIOR TO COMPLETION OF CAUSE OF DEATH? ☒YES ☐NO

29a. Was decedent pregnant at time of death? ☐YES ☒NO ☐UNKNOWN

29b. Was decedent pregnant during the last 12 months? ☐YES ☒NO ☐UNKNOWN

Figure 8-36 *Death certificate.*

The gyral tips of the brain (Figures 8-32 and 8-37) are located over sharp and bony protuberances within the skull. Thus, this region of the brain frequently is bruised by a head "jostling," regardless of direction of impact or head motion. Thus, although it may seem counterintuitive, this injury is not specific as far as indicating a frontal head impact. In fact, it is quite consistent with the "no significant head impact" theory (Pike, 1990).

Figure 8-38 is not a record from this case but has been added to provide additional background regarding brain injury. As already mentioned, if sufficient pressure is exerted on the brain by intracranial fluid collection, the brain will be distorted and, in severe instances, actually will be displaced through one of the openings in the structure surrounding the brain. In this instance, the records refer to brain deformation around the edge of the tentorium (Figure 8-38, Label T)—a somewhat horizontal membrane that "fits" in a groove in the brain structure and helps to hold the brain in place during head motion. The brain's tonsilar region also was affected, with its displacement possibly involving the foramen magnum (Figure 8-38, Label F), the large opening at the base of the skull.

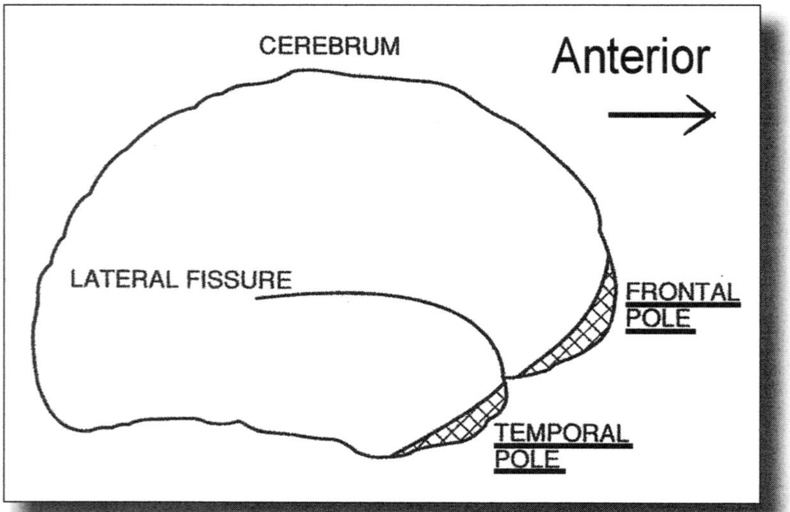

Figure 8-37 *Sites of frequent brain contusion (From Pike, 1990, Figure 2-6. Published by SAE International. Used with permission.)*

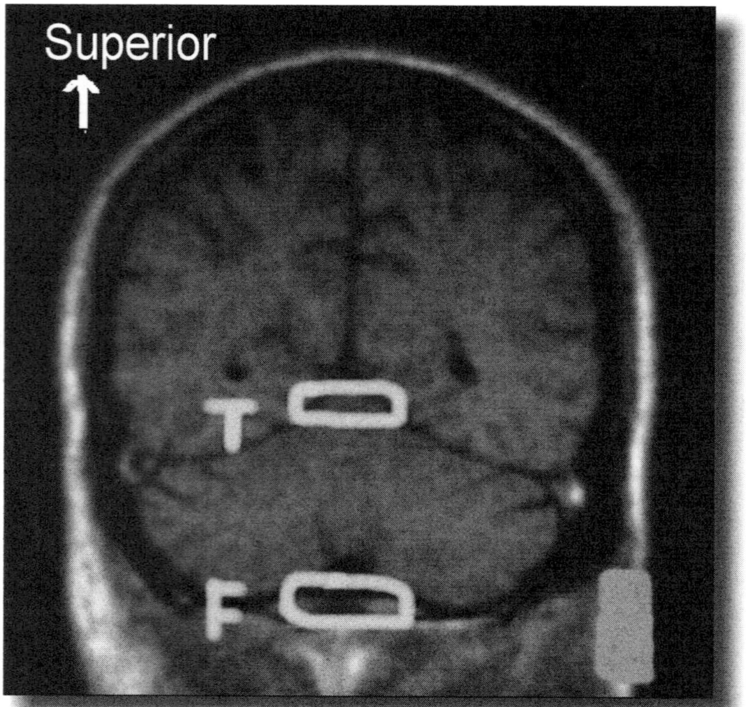

Figure 8-38 *Brain MRI, posterior coronal view, herniations.*

Summary

In summary, this is what I consider to be the most likely scenario. This patient was the driver of a vehicle that was involved in essentially a high-speed rear impact. Although the driver initially was moved rearward (i.e., toward the point of impact), there was enough energy to have the driver rebound and move forward. The forward motion of the unbelted driver produced knee impact with the lower instrument panel. The driver's forward motion also produced chest impact with the steering wheel and/or instrument panel. This, in turn, caused chest compression, rib fracture, and an abrupt stopping of the torso's forward motion. In turn, stopping of the torso produced neck flexion that produced a vertebral fracture and chin bruise. Thus, the chin bruise was not indicative of an "upper-cut" but rather, in this instance, was produced when the neck flexed and the head moved forward and struck the steering assembly or other structure. The combination of a violent rearward head motion followed by forward head motion produced sufficient brain tissue trauma, even without significant impact between the outer surface of the head and some part of the vehicle interior, to initiate a cascade that included a progressive edema and an intracranial pressure that was not compatible with life.

Chapter 9

א

Advanced Case Study III: Parking Lot

Introduction

Similar to the other advanced case studies, this study provides a number of records (Figures 9-1 through 9-8). The goal is to review them and to integrate them. However, this advanced case study is different from the others in three main ways. First, a summary of some of the background information is provided, rather than the actual records (e.g., Figure 9-2). Second, three possible mechanisms are identified, and readers are requested to associate one or more of the three with each of the significant injuries and then to justify the choice. Third, readers are asked to address if the cord injury occurred all at once, or if it occurred over time and is attributable, at least in part, to events subsequent to an initial cord injury.

The records are arranged as follows: current scenario (Figure 9-1); composite medical history (Figure 9-2); radiology report for the head, neck, and chest (Figure 9-3); radiology report for the back and chest (Figure 9-4); radiology report for the chest (Figure 9-5); cervical spine film image (Figure 9-6); radiology report for the cervical spine (Figure 9-7); and consultation letter (Figure 9-8).

A discussion follows Figure 9-8. After that discussion, the first eight figures are repeated (Figures 9-9 through 9-16) but are supplemented with additional background material. Also, the repeated figures include some callouts and/or marginalia that have been added to the original figures. Figure 9-9 is an annotated version of Figure 9-1, Figure 9-10 is annotated version of Figure 9-2, and so forth (with the exception of Figure 9-8, which appears as Figure 9-17 to allow the inclusion of a supplemental new figure as Figure 9-16). For those readers who wish to try the analysis on their own first, it is recommended that they not look past Figure 9-8 until they have tried the exercise. For those who prefer another "guided tour," it is recommended that they go directly to the Discussion section and the second set of figures.

Current Scenario:

Patient went out for the evening.

During evening experienced personal violence (involved in a fight), slip and fall and vehicle impact (as pedestrian.) Patient found, injured, on floor of parking lot.

Patient helped in to private vehicle and taken home. She was not fully coherent during ride. She may have been complaining of tingling feeling in her hands.

Pt intoxicated, not sure of many details.

Pt helped into house, lay down on floor and did not want to be moved.

No treatment overnight.

Next morning, taken to hospital.

Cord injury at C6 (site of previous fusion), quadriparesis.

Figure 9-1 *Current scenario.*

Composite Medical History

25 years ago:
Whiplash → neck pain

8 years ago:
Burning in neck, radiated into right hand and shoulder
→ discectomy & fusion, C5/C6

Figure 9-2 *Previous medical history.*

3 VIEW SKULL: 3 views of the skull show no definite fractures. Density over the superior right orbit apparently is on the surface of the patient. Medial wall soft tissue density in the left antrum probably retention cyst.

LATERAL CERVICAL SPINE POST HALO: Impaction of C5-6 with spinous process fracture again evident. The amount of posterior angulation has not appreciably changed.

PRONE FILM LATERAL CERVICAL SPINE: No change.

CHEST: Right scapular and multiple right sided rib fractures. Alveolar infiltrate right upper lobe due to contusion as well as a small amount of pleural fluid on the right unchanged.

Figure 9-3 *Radiology report for the head, neck, and chest.*

DEPARTMENT OF NEURORADIOLOGY/SCANNING

Age:

Dr. ——— X-RAYS Room No:

DORSAL SPINE: The tenth rib is shown to be fractured on the lateral view. The ninth rib on the right posteriorly is also fractured on the AP dorsal spine film. There is mild dorsal scoliosis convex to the left. No other dorsal spine fracture evident.

LUMBAR SPINE: Mild lumbar scoliosis convex to the left. Lumbar spine appears intact with no fracture. Psoas margins are well delineated.

PELVIS: Negative except for defect left ilium related to donor site over the anterior interbody fusion.

CHEST: Mild to moderate elevation of the right diaphragm. There is a small amount of residual alveolar infiltrate in the right upper lung zone which represents a significant improvement over the prior study. The retrocardiac infiltrate on the left is now largely cleared. Multiple rib fractures which appear to be on the right now appear to involve the fourth, fifth, sixth, seventh, eighth and ninth ribs.

Figure 9-4 *Radiology report for the back and chest.*

DEPARTMENT OF NEURORADIOLOGY/SCANNING

Age:

Dr. X-RAYS Room No:

CHEST: Chest shows a right fourth rib fracture with probable fifth and seventh rib fractures on the right with pleural fluid and moderate alveolar infiltrate in the right lung consistent with pulmonary contusion. There is an air bronchogram in the left retrocardiac region consistent with parenchymal infiltrative change which could even represent aspiration. There is shallow inspiration. NG tube is in the area of the stomach. Pulmonary vascular congestion is present as well. The chest represents a deterioration as opposed to the outside film with poor inspiration and a greater degree of alveolar infiltrative changes.

Figure 9-5 *Radiology report for the chest.*

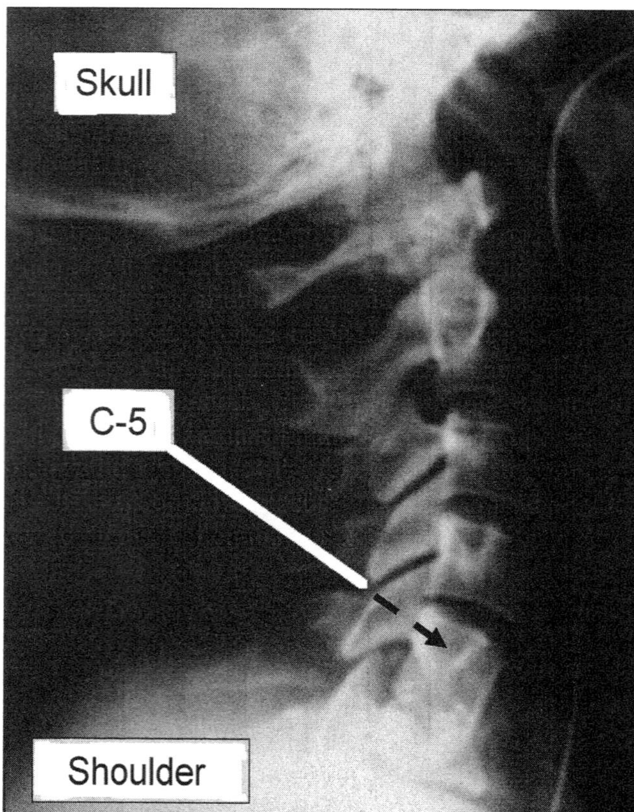

Figure 9-6 *Cervical spine film image.*

CERVICAL SPINE: The cervical spine shows evidence of previous anterior interbody fusion of C5-6, however there appears to be a traumatic impaction of the C5-6 vertebra at the fusion site with compression of central and anterior aspect of C6 with a small fragment of bone seen anteriorly in conjunction with a spinous process fracture which is really evulsed and separated by 1 centimeter. The distal fragment being displaced caudally. There also appears to be a probable fracture of facet and a question of pedicle fracture on the right although that is not definite. The superior articulating facet of C6, however, does show a fracture at the tip. By plain film the sagittal diameter of the canal, however, does not appear to be compromised. The pillar on the right at C5 is compressed. There is a moderate amount of posterior angulation at the C5-6 level with the disc space being obliterated from the previous interbody fusion.

Figure 9-7 *Radiology report for the cervical spine.*

the patient did have a concussion at the time of the accident based upon her amnesia for the events and also because she was reported to be unconscious. She did have black eyes, which were separate injuries from her neck injury.

Figure 9-8 *Consultation letter.*

Note: Immediately following this note is a discussion of the case, followed by a repeat of the first eight figures. However, this second presentation of the figures includes callouts and/or marginalia, as well as some additional figures with background material. For those readers who wish to try the analysis on their own first, it is recommended that they not read the Discussion section until they have tried the exercise. For those who prefer another "guided tour," it is recommended that they go directly to the Discussion section and the second set of figures.

Discussion

The first document (Figure 9-9), titled "Current Scenario," is a compilation of information regarding the current trauma. Although this is not an actual record, the reader should consider this document as part of his or her analysis. This document indicates that the patient was found injured on the floor of a parking lot. The patient remembers going out for the evening

Current Scenario:

Patient went out for the evening.

During evening experienced personal violence (involved in a fight), slip and fall and vehicle impact (as pedestrian.) Patient found, injured, on floor of parking lot.

Patient helped in to private vehicle and taken home. She was not fully coherent during ride. She may have been complaining of tingling feeling in her hands.

Pt intoxicated, not sure of many details.

Pt helped into house, lay down on floor and did not want to be moved.

No treatment overnight.

Next morning, taken to hospital.

Cord injury at C6 (site of previous fusion), quadriparesis.

Figure 9-9 *Current scenario.*

but cannot remember how she wound up injured and on the floor of a parking lot. The patient was intoxicated and presented a very sketchy account of the evening. It appears that during the evening, she was involved in a fight, a slip and fall, and a vehicle impact (as a pedestrian). The reader is asked to study the available material and to assess the following: (1) which event(s) produced the paralyzing spinal injury, and (2) if the patient might have benefited from earlier treatment. Of course, there is nothing to preclude the reader from attributing a given injury to a combination of two or three scenarios.

This first document (Figure 9-9) also includes the information that the patient previously had a fusion of the lower cervical spine and that the current scenario produced cord injury at the site of the previous fusion.

The next document (Figure 9-10), titled "Composite Medical History," likewise is a compilation of several sources regarding the current trauma. Although this is not an actual record, the reader should consider this document as part of his or her analysis. As is often the case with verbal histories covering previous trauma, the recall is not precise or detailed. It indicates that the patient had a whiplash-type injury approximately 25 years ago. Approximately eight years ago, the patient had neck and upper right extremity symptoms that resulted in her being treated with a discectomy and spinal fusion at C5/C6. That brings the record up to date, and the material then resumes with records from the current scenario.

Composite Medical History

25 years ago:
Whiplash → neck pain

8 years ago:
Burning in neck, radiated into right hand and shoulder
→ discectomy & fusion, C5/C6

Figure 9-10 *Previous medical history.*

The radiology report documents (Figures 9-11 through 9-13) describe a number of injuries, including the following: right scapula fracture, right rib fractures (ribs 4–9), right lung contusion, impaction of C5/C6 vertebrae, and spinous process fracture. The scapula and rib fractures and lung contusion all appear consistent with a compression of the shoulder region by a slow-moving vehicle. There are no lower limb injuries that would point toward a pedestrian impact; thus, the scenario appears to be a run-over of the right shoulder region after the patient was on the ground. This document also notes that the injury produced quadriparesis, a partial paralysis affecting all four limbs.

The cervical spine x-ray (Figure 9-14) in conjunction with the radiology report (Figure 9-15) provide additional detail of the spine, namely, traumatic impaction of C5/C6 vertebrae at the site of a previous fusion, compression of the central and anterior aspect of C6 as well as facet fractures at C5 and C6, and an avulsed spinous process fracture. The radiologist also notes (Figure 9-15) that the diameter of the canal does not appear to be compromised (an observation which, at least initially, would provide some basis for optimism regarding the preservation of spinal cord function).

Although the spinous process may have been fractured by a vehicle impact or by personal violence directed to the back of the patient's neck, there are no surface injuries to support this. Rather, the avulsed nature of the spinous process fracture provides a basis for

3 VIEW SKULL: 3 views of the skull show no definite fractures. Density over the superior right orbit apparently is on the surface of the patient. Medial wall soft tissue density in the left antrum probably retention cyst.

LATERAL CERVICAL SPINE POST HALO: Impaction of C5-6 with spinous process fracture again evident. The amount of posterior angulation has not appreciably changed.

PRONE FILM LATERAL CERVICAL SPINE: No change.

CHEST: Right scapular and multiple right sided rib fractures. Alveolar infiltrate right upper lobe due to contusion as well as a small amount of pleural fluid on the right unchanged.

Figure 9-11 *Radiology report for the head, neck, and chest.*

DEPARTMENT OF NEURORADIOLOGY/SCANNING

Age:

Dr. ——— X-RAYS Room No:

DORSAL SPINE: The tenth rib is shown to be fractured on the lateral view. The ninth rib on the right posteriorly is also fractured on the AP dorsal spine film. There is mild dorsal scoliosis convex to the left. No other dorsal spine fracture evident.

LUMBAR SPINE: Mild lumbar scoliosis convex to the left. Lumbar spine appears intact with no fracture. Psoas margins are well delineated.

PELVIS: Negative except for defect left ilium related to donor site over the anterior interbody fusion.

CHEST: Mild to moderate elevation of the right diaphragm. There is a small amount of residual alveolar infiltrate in the right upper lung zone which represents a significant improvement over the prior study. The retrocardiac infiltrate on the left is now largely cleared. Multiple rib fractures which appear to be on the right now appear to involve the fourth, fifth, sixth, seventh, eighth and ninth ribs.

Figure 9-12 *Radiology report for the back and chest.*

DEPARTMENT OF NEURORADIOLOGY/SCANNING

Age:

Dr. X-RAYS Room No:

CHEST: Chest shows a right fourth rib fracture with probable fifth and seventh rib fractures on the right with pleural fluid and moderate alveolar infiltrate in the right lung consistent with pulmonary contusion. There is an air bronchogram in the left retrocardiac region consistent with parenchymal infiltrative change which could even represent aspiration. There is shallow inspiration. NG tube is in the area of the stomach. Pulmonary vascular congestion is present as well. The chest represents a deterioration as opposed to the outside film with poor inspiration and a greater degree of alveolar infiltrative changes.

Figure 9-13 *Radiology report for the chest.*

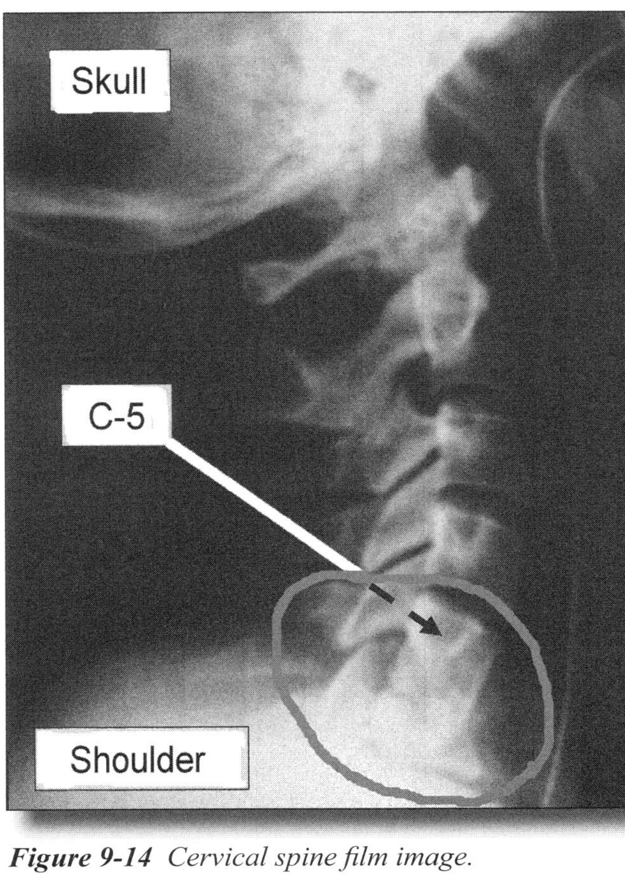

Figure 9-14 *Cervical spine film image.*

CERVICAL SPINE: The cervical spine shows evidence of previous anterior interbody fusion of C5-6, however there appears to be a traumatic impaction of the C5-6 vertebra at the fusion site with compression of central and anterior aspect of C6 with a small fragment of bone seen anteriorly in conjunction with a spinous process fracture which is really evulsed and separated by 1 centimeter. The distal fragment being displaced caudally. There also appears to be a probable fracture of facet and a question of pedicle fracture on the right although that is not definite. The superior articulating facet of C6, however, does show a fracture at the tip. By plain film the sagittal diameter of the canal, however, does not appear to be compromised. The pillar on the right at C5 is compressed. There is a moderate amount of posterior angulation at the C5-6 level with the disc space being obliterated from the previous interbody fusion.

Figure 9-15 *Radiology report for the cervical spine.*

classifying it as a so-called "clay shoveler's" fracture. This term applies to an avulsion fracture of the lower cervical spine (as is the present case), that is, the spinous process was broken off by the ligamentous force exerted on the process during a violent flexion (Harris, 1989). In my opinion, the sudden flexion was induced by personal violence. Figure 9-9 states that the patient was helped into a private vehicle (the extent of the help is not clear), so it is difficult to know her level of lower limb function at that time. During the ride to her house, she may have complained of a tingling sensation in her hands; if this were the case, it would support the notion that she still had some mobility in her lower limbs at a point in time that she was having neurological symptoms involving her upper limbs.

The patient had difficulty walking, but we do not know how much of this at the time was due to intoxication and how much was due to cord injury. Although typical cord injury produces cumulative distal loss, that is, if the primary injury is at a given level, functional and sensory loss also will affect body regions innervated by the segment of the cord distal to the injury site (i.e., affect the function of parts of the body that are located below the injury site). However, this need not be the case in the event of a "central cord syndrome," wherein the main deficits involve the upper limbs but the lower limbs are relatively unaffected (Pike, 1990). The tracts of the spinal cord have a laminated structure (Figure 9-16); therefore, injury to the center of the cord tends to affect the "inside tracts," that is, the layers that are located radially inward and that innervate the upper limbs. When this occurs, the effects typically are to the upper limbs, with relatively little effect on the "outside" tracts that go to the lower limbs. The central cord syndrome is more frequent in older individuals who already have degenerative changes (e.g., bulging discs and buckling ligaments that reduce the space in the spinal canal and set the stage for the cord to be compressed both front and back). Although this individual was not elderly, the cervical region of her spine previously had been injured. Figure 9-9 indicates that

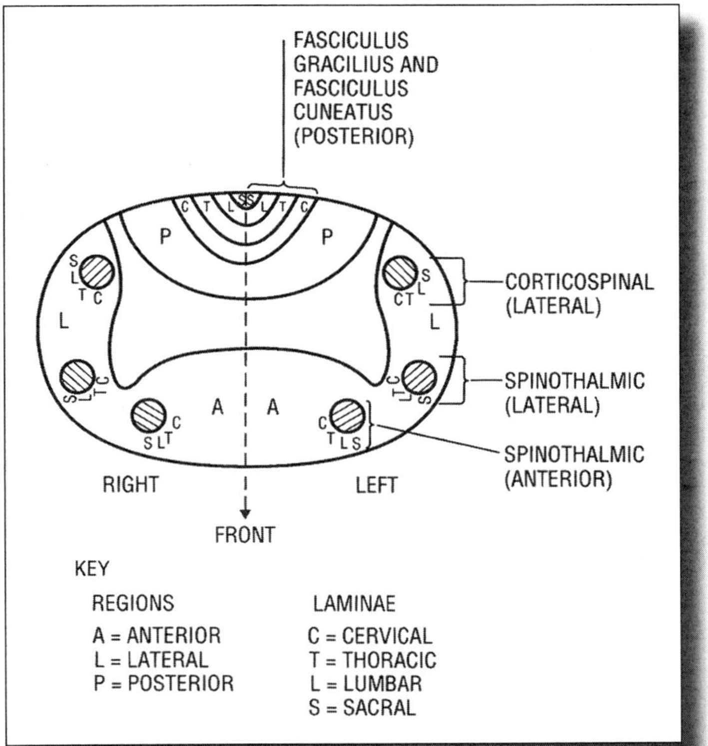

Figure 9-16 *Lamination of the spinal cord tracts. (From Pike, 1990, Figure 2-14. Published by SAE International. Used with permission.)*

the patient experienced upper limb symptoms. If her difficulty in walking was due primarily to intoxication, then this may be an instance of central cord syndrome. If the difficulty walking was due primarily to cord injury, then this was not a central cord type of phenomenon.

A letter from a consulting physician (Figure 9-17) states that the patient had a concussion and offers as a basis that she has amnesia for the events and because she was reported to be

> the patient did have a concussion at the time of the accident based upon her amnesia for the events and also because she was reported to be unconscious. She did have black eyes, which were separate injuries from her neck injury.

Figure 9-17 *Consultation letter.*

unconscious. It is not known if this physician was aware of the patient's intoxication at the time (which also could be a reason for her amnesia for the events and, indeed, for her being unconscious). Furthermore, it is not known if the physician was made aware of the intoxication of the person who brought her home, which would affect the credibility of any history that person may have provided.

It generally is accepted that alcohol may affect injury outcome on a cellular level (Fry, 1987). In this instance, there may have been the added alcohol effects of being less able to determine the need for help, less able to express the need for help, and the possible masking of central nervous system (CNS) symptoms. Furthermore, alcohol may have affected the judgment of the patient's roommate regarding assessing her condition and whether or not she needed emergency care. Note that although a spinal injury is not paralyzing per se, if not treated, paralysis can result. For example, the initial injury may tear spinal ligaments that hold the vertebrae in place, and the resulting hypermobile vertebrae may induce cord damage simply from everyday neck movements. Also note that any movement of the extremities, such as may have been observed when the patient was being brought home, can be an indicator that the outcome may have been mitigated by early treatment (Baxt, 1985).

Summary

There is much uncertainty regarding these events, but my assessment is that the most likely scenario was a combination of the following: personal violence that produced the black eyes and sudden neck flexion, a slip and fall, stumble and fall, or additional personal violence that positioned the patient on the parking lot floor, and then a vehicle hit and run, with the vehicle running over the patient's right shoulder region. Although no tire-like abrasions were noted to the patient's shoulder region, heavy clothing may have protected her from these. The records provided do not address the patient's outer clothing, the time of year, or the climate. In fact, if the patient had been wearing summer clothing, I would conclude that she was struck in the shoulder/upper chest region by a person (perhaps using a club-like object) or vehicle, and the combination of this and her subsequent fall produced the rib fractures and pulmonary contusion. In any event, it appears that the combination of fighting, falling, and possibly being struck or run over by a vehicle produced a cord injury at the site of a previous fusion and that this was exacerbated by delayed treatment. Although not addressed in the records provided for this study, there also is the possibility of a subsequent fall after the patient was brought home and before she received medical attention.

Chapter 10

א

Examples

This chapter provides several sample reports. The reports (based on hypothetical events and/or real events with names, dates, and other potential identifiers modified) provide examples of the types of issues that biomechanics consultants may address, a general idea of the types of information they utilize in coming to their opinions, and a format that is used to report on their work. Note that in some instances, a biomechanic may be asked to determine overall impressions, such as "How did the injuries occur?" In other instances, he or she may be asked to address one or two very specific topics, such as "How much of their current complaint is due to the subject incident?" or "How would the injuries have been affected if the occupant had been wearing his or her seat belt?" The examples that follow provide one perspective on each of the events. (An example of two perspectives on the same event is provided in Chapter 3, Figures 3-7 and 3-8.)

Example 1

In this example, a vehicle was impacted at the driver's side door, and the driver was complaining of temporomandibular joint (TMJ) pain. The situation is complicated by the fact that the driver had a history of TMJ injury, and one important aspect of the analysis is to address the contribution of the previous injury. Note that some of the most relevant information is "indirect," that is, the cranial nerve testing. Although this was not a test of the TMJ per se, the testing involved contact with and motion of the TMJ.

CONSULTATION REPORT

ABC Consulting, Inc.
12345 Main Street
Anytown, MI 48111
Phone: (313) 123-4567
Fax: (313) 123-5678
Expert1@ABCconsult.com

To:	A. Lawyer, Esq. Able, Baker, and Delta, LLP
Fax:	(111) 123-4567
From:	Edward Expert
Date:	January 1, 2010
Re:	Jane Doe
DOI:	February 29, 2009
YOB:	1950

[Note: "DOI" refers to the date of the incident, and "YOB" is the subject's year of birth.]

Per your request, I have reviewed the documents you provided, including medical records, various x-rays and MRIs [magnetic resonance imaging], and a police accident report. My preliminary comments, which follow, may be revised as additional information is provided.

OPINIONS

1. There was no violent head motion, no head impact, no loss of consciousness, and no biomechanical basis for head injury.

2. There was no violent neck motion, no impact to the jaw or TMJ region, and no biomechanical basis for TMJ injury.

3. The TMJ injury was pre-existing and was not causally related, biomechanically, to the 2009 incident.

BACKGROUND

Ms. Jane Doe was the driver of a 2008 sedan that had an impact with a 2007 van driven by Mr. John Smith. The impact involved primarily the right front of the van and the left front (driver's-side) door area of the sedan. Ms. Doe was injured in a previous motor vehicle accident in 2005 (hereafter referred to as the "2005 incident"), which produced various injuries, including TMJ injury.

DISCUSSION

The medical records indicate that Ms. Doe had a TMJ injury in 2005, for which she received treatment during 2005 through 2008. The records indicate her last office visit for TMJ treatment was September 8, 2008. She was to return for a follow-up visit on December 12, 2008 but cancelled that appointment and did not reschedule between then and the time of the subject incident. Thus, her treatment for the 2005 TMJ injury appears to have been interrupted rather than completed at the time of the subject incident.

With regard to the 2009 incident, Ms. Doe reported no TMJ pain at the accident scene. Also, the ER [emergency room] records indicate that her cranial nerves were tested, and this provides indirect but relevant TMJ information—the cranial nerve testing includes moving the jaw and examiner contact with the jaw and the TMJ region and again, this movement and contact did not evoke any report of scalp, skull, jaw, or TMJ pain. The fact that there was no skull pain and no pain on palpation of the skull also serves as indicators that there was no significant head impact. The ER records also indicated no loss of consciousness at the scene and that the patient was alert in the ER. Thus, these records point to no violent head motion and hence no violent neck motion, with or without head impact. An electroencephalogram (EEG) taken in March 2009 indicated the patient's brain was normal, and a brain MRI the same month also was normal. A cervical spine MRI from April 2009 indicated some subtle degenerative changes in the C6/C7 region. These degenerative changes would be consistent with Ms. Doe's age and the 2005 incident.

Ms. Doe reported wearing her seat belt. The medical records reported no chest tenderness and no chest bruising. This indicates that the shoulder belt was not required to exert substantial restraint force to these areas. Also, the abdomen and hips were not bruised and not tender, indicating there was no substantial restraint force from the lap belt. The vehicle damage was primarily superficial, involving sheet metal and other "soft" structures that do not require much energy to deform. Thus, vehicle deformation indicated a vehicle impact that was of relatively low severity. The police report indicated that the (frontal) airbags in neither vehicle deployed. The airbag systems monitor the movement of the vehicle and are designed to deploy in moderate or severe frontal impacts. The non-deployment in both vehicles is an indicator that the impact was in the "sub-deployment" or relatively non-severe range. The low-speed nature of the impact, in conjunction with the lack of occupant bruising from seat belt restraint forces, and the lack of occupant injury from vehicle interior contact are consistent in indicating that the occupant's movement during the vehicle impact was rather benign; such movement would not be expected to produce long-term injury sequelae.

Thank you for this consultation.

Example 2

This example involves an impact between a passenger car and a van, with the van impacting the passenger car in the vicinity of the car's right front passenger door. The right front passenger sustained skeletal and soft-tissue injuries, and the consultant was asked to address two topics: (1) How would the patient's injuries have been affected if the impact speed had been 1 to 5 mph less, and (2) How would the patient's injuries have been affected if she had been wearing her seat belt?

CONSULTATION REPORT

ABC Consulting, Inc.
1234 Main Street
Anytown, MI 48888
Phone: (313) 123-4567
Fax: (313) 234-5678
expert1@abc.com

To: A. Lawyer, Esq.

Phone: (123) 456-7890

Fax: (123) 456-7891

From: D. Expert, Ph.D.

Date: February 30, 2010

Re: Jane Doe

DOI: December 1, 2009

YOB: 1977

OPINIONS

1. The injuries to plaintiff Jane Doe's right side, including fractures of the acetabulum and sacrum, were produced when the van struck the right side of the sedan.

2. If the pre-braking speed of the van had been 55 mph instead of 56–60 mph, Ms. Doe's injuries may or may not have been mitigated.

3. The injuries to Ms. Doe's left side, including rib fracture and kidney laceration, were produced when the left side of her body impacted objects and/or structure within the vehicle.

4. The left-side injuries mentioned in Opinion 3 would have been mitigated or eliminated if Ms. Doe had been wearing her seat belt.

The opinions are stated to a reasonable degree of biomechanical certainty. My experience and qualifications are outlined in my attached CV. The comments are based on the material listed here and may be revised if additional material becomes available.

MATERIAL

Police investigation report, 12/2/09

Deposition of Officer Paul Jones

Hospital Medical Center (records for Jane Doe)

Hospital Medical Center (records for Alice Jones)

Vehicle repair estimate, Ace Body Shop

Traffic citation (unsafe turn)

Acme Rehabilitation Center, records for Jane Doe

Police accident report (12/1/09)

BACKGROUND

This is a two-vehicle impact, involving a 2008 four-door sedan and a 1997 van. The van was driven by John Smith, and the sedan was driven by Alice Jones. The sedan also had a right front passenger, Jane Doe. The impact primarily involved the front of the van and the right side of the sedan.

The speed of the van was calculated to be 56–62 mph at the start of pre-impact braking and 42–48 mph at impact (Consultant Report 1). (Note: The 1997 van was not equipped with an electronic data recorder [EDR] that recorded the vehicle speed. EDR-equipped passenger vehicles are scheduled to have recording capability governed by federal regulation beginning with the 2011 model year.) Both front airbags in the van deployed. The sedan was equipped with a driver's-side-only airbag, which did not deploy. At the time of the impact, the van was going forward, and the sedan was stopped.

Mr. Smith reported no injury. Ms. Jones reported initially feeling somewhat dazed, but she soon reported feeling better and did not seek medical attention. Ms. Doe received multiple fractures of the pelvic girdle, as well as soft-tissue injuries. The repair estimate for the sedan was $7,610.50.

DISCUSSION

The subject vehicle was impacted on the right side, close to Ms. Doe's seating position. The impact pushed the vehicle side structure into Ms. Doe's right side, producing a variety of skeletal and soft-tissue injuries. Ms. Doe was not restrained; therefore, the right side impact also imparted a sideways velocity to her. She was projected toward the driver's side of the vehicle impacting her left side and impacting the driver, who was seat belted in place. The initial impact to her right side produced a number of injuries, including multiple fractures of the right pelvis, and her subsequent impact with structure/objects toward the center or left of the vehicle, produced her left side injuries, including the rib fractures and kidney laceration, and may have contributed to her left acetabular fracture.

You have asked me to address the effect, if any, on Ms. Doe's injuries, if the pre-braking speed of the car had been 55 mph compared to the calculated range of 56–60 mph. The abbreviated injury scale (AIS)[(1)] is a widely used and well-established methodology[(2)] for scaling injury severity. The AIS groups injury into one of six levels, and the injuries within a given level are considered to be comparable. [(3)] Thus, although an impact speed of 1–5 mph less would be somewhat less severe, it still may be in the range of impact severity that would have produced a similar injury pattern. Although one can make a general, qualitative statement that the lower the impact speed, the less severe the impact, there is not a well-established methodology that would permit a precise, quantitative prediction of the specific injuries that would have been sustained by Ms. Doe if the impact speed were reduced by a specified amount.

In this instance, the pelvic injuries included fracture fragments and bony displacement (as opposed to bone "cracks"). Therefore, it is unlikely that this impact severity was just at the level of fracture threshold.[(4)] Rather, it is most likely that similar fractures would have resulted even if the impact severity had been slightly less. It appears that Ms. Doe's left side injuries, including the rib fractures and kidney laceration, were caused by left side impact with the vehicle interior and/or interior objects. Seat belts have been shown to mitigate near-side occupant injury in side impacts,[5] and if Ms. Doe would have been properly wearing the available lap-shoulder belt, her movement would have been limited, her deceleration would have been more gradual, and her left side impact injuries would have been mitigated or eliminated.

Thank you for this consultation.

REFERENCES

1. Abbreviated Injury Scale (AIS), Association for the Advancement of Automotive Medicine, Barrington, IL, 1985.

2. Federal Motor Vehicle Safety Standard 214, Side Impact Protection, National Highway Traffic Safety Administration, Washington, DC, 2004.

3. Pike, J.A., *Automotive Safety: Anatomy, Injury, Testing and Regulation*, Society of Automotive Engineers, Warrendale, PA, 1990, 1997.

4. Levine, R., "Injury to the Extremities," in Nahum, A.M., *et al.*, eds., *Accidental Injury: Biomechanics and Prevention*, Springer-Verlag, New York, 1993.

5. Ranianthra, J.N., "Can Improvements Be Made in Automobile Seat-Belt Technology," 84th Annual Meeting of the Transportation Research Board, Jan. 12, 2005, Transportation Research Board, Washington, DC, 2005.

Example 3

The basic scenario for Example 3 is that Mr. Smith has a history of shoulder trauma. Then, a few months after a very minor vehicle impact, he was found to need rotator cuff surgery. The consultant was asked to address two questions: (1) Are the shoulder conditions attributable to the most recent motor vehicle incident, and (2) if not, is there another cause that can be identified?

CONSULTATION REPORT

ABCD Consulting, Inc.
12 Main Street
Anytown, MI 48888
Phone: (313) 123-4567
Fax: (313) 234-5678
expert1@abcd.com

To:	Ann Attorney, Esq. ABC Law
E-mail:	aattorney@abclaw.com
Phone:	(123) 456-7890
Fax:	(123) 456-7891
From:	D. Expert
Date:	November 31, 2010
Driver:	Stanley Smith
DOI:	July 9, 2009
YOB:	1956

OPINIONS

1. The subject vehicle impact was very minor.
2. The right shoulder Type III acromion, rotator cuff tear, bicipital tendon tear, and superior labral tear are not causally related, biomechanically, to the subject motor vehicle incident.

The opinions are stated to a reasonable degree of biomechanical certainty. My experience and qualifications are outlined in my attached CV. The comments are based on the material listed here and may be revised if additional material becomes available.

MATERIAL

Medical records, including the following:

Main Street Hospital

Western Rehabilitation Center/A. Baker, M.D.

S. Delta, M.D.

Traffic accident reports (2005 and 2009 incidents)

Deposition of S. Smith, February 13, 2010

BACKGROUND

Stanley Smith was the driver of a 2008 sedan when the right front of the vehicle impacted the base of an 11-foot-high statue. The impact produced minimal damage to the subject vehicle (total repair estimate less than $800). Mr. Smith was not wearing his seat belt, the airbag did not deploy, and no other vehicle was involved. Mr. Smith declined medical attention at the scene and was able to drive the vehicle from the accident scene. Three days later, he went to a hospital emergency room, complaining of bilateral shoulder pain. He was examined and x-rayed. The x-rays were read as not indicating any injury, and Mr. Smith was sent home.

HISTORY

The history of the driver of the subject vehicle includes the following:

1992 Started current occupation of painter.

2004 Slipped and fell on ice; hurt both wrists and both shoulders; received left rotator cuff surgery.

2005 A vehicle driven by Mr. Smith had a head-on collision with an SUV. Smith's vehicle airbag deployed. Smith's vehicle repair estimate was slightly more than $6,000. Right shoulder complaints included pain and limited movement.

2009 Subject incident.

DISCUSSION

An MRI of the patient's right shoulder (on September 31, 2009) indicated the following: a Type III acromion and degenerative changes at the supraspinatous tendon. The Type III acromion is part of the shoulder structure and is not attributable to the subject incident. The Type III structure provides reduced space available for tendon movement during shoulder abduction and thereby makes the person prone to rotator cuff injury over a period of time. [1, 2] (The rotator cuff is a group of four muscles and their associated tendons that help to hold the shoulder in place and help to perform some shoulder movements, including raising the arm to the side.) The surgeon's report provided a diagnosis of right shoulder rotator cuff tear. The surgical report noted "significant degenerative" changes. The degenerative conditions noted on the MRI and surgical reports would have occurred over time and biomechanically are not attributable to the subject motor vehicle impact.

The subject incident was of minor severity, as evidenced by the lack of airbag deployment and the minimal vehicle damage. The airbag system monitors vehicle movement and uses it to calculate if the airbag needs to deploy. In this instance, the impact was too minor to warrant airbag deployment. Furthermore, the vehicle damage was minimal and was confined primarily to sheet metal and plastic around the driver's side bumper and headlight region of the vehicle. These are relatively "soft" structures and do not require a great deal of energy to deform.

A widely used and well-established[3, 4] measure of crash severity is "delta v," a number representing the speed at which a reference crash would produce similar vehicle damage. (That is, if an exemplar vehicle were to impact a solid wall at a speed of delta v, it would replicate the vehicle damage that occurred in the original impact.) Airbag systems monitor the movement of a vehicle, and, in this impact scenario, typically would deploy if the delta v was in the range of approximately 10–15 mph delta v. The non-deployment would be an indication that the impact was in the "sub-deploy" level of severity. More specifically, based on vehicle deformation, the damage due to the subject impact was minor and was consistent with an impact of less than 5 mph delta v. This is quite comparable to the delta v produced in amusement park bumper car impacts.[5] Note that bumper car occupants are of a wide range of ages and frailty and typically experience repeated impacts, from multiple directions, without injury. Thus, vehicle impacts of severity comparable to this incident are readily withstood by children and the elderly.

In summary: (1) the subject incident involved a vehicle impact of very minor severity and would not be expected to injure an adult who was properly lap- and shoulder-belted in the driver's seating position, (2) the right shoulder conditions that were treated surgically were of a congenital/degenerative nature and were not attributable to the subject incident, and (3) the right shoulder conditions that were treated surgically are attributable to the combination of the shape of the acromion, two slip and fall incidents, the 2005 severe motor vehicle crash, and the patient's occupation as a painter (which requires frequent, overhead raising of the arm and hence frequent rotator cuff usage).

Thank you for this consultation.

REFERENCES

1. Jobe, C.M., "Gross Anatomy of the Shoulder," in Rockwood, C.A., *et al.*, *The Shoulder*, 2nd Ed., Vol. 1, W.B. Saunders, Philadelphia, 1998.

2. Steinbach, L.S., *et al.*, *Shoulder Magnetic Resonance Imaging*, Lippincott, Philadelphia, 1998.

3. Fricke, L.B., *et al.*, *Traffic Accident Investigation Manual*, Northwestern University, Evanston, 1990.

4. Supplemental Notice of Proposed Rulemaking, Docket No. NHTSA 99-6407; Notice 1, FMVSS 208, Occupant Crash Protection, National Highway Traffic Safety Administration, Washington, DC, 1999.

5. Siegmund, G.P., *et al.*, "Speed Change (Δv) of Amusement Park Bumper Cars," in *Proceedings of the Canadian Multidisciplinary Road Safety Conference*, Saskatoon, June 14–16, 1993.

Example 4

This example involves a low-speed rear impact. The consultant addressed the driver complaints of whiplash injury and the x-ray findings of neck straightening.

CONSULTATION REPORT

ABC Consulting, Inc.
1234 Main Street
Anytown, MI 48888
Phone: (313) 123-4567
Fax: (313) 234-5678
expert1@abc.com

To: Ann Attorney, Esq.

E-mail: aattorney@lawfirm.com

From: D. Expert

Date: June 31, 2010

Subject: Joe Smith

DOI: September 31, 2009

YOB: 1957

As you requested, I have reviewed the material that you provided, primarily medical records, photographs, and repair estimates of the two vehicles and a report by A. Beta. The following comments are based on that material and may be revised if additional material becomes available. The opinions are stated to a reasonable degree of biomechanical certainty.

OPINIONS

1. This MVA (motor vehicle accident), with regard to the Smith vehicle, was primarily a minor rear impact.
2. This MVA does not provide a biomechanical basis for long-term injury sequelae.
3. Mr. Smith's complaints are not causally related, biomechanically, to the subject MVA.

BACKGROUND

The plaintiff, Joe Smith, was the driver of a 2010 sedan that was involved in an impact with another passenger car, a 2009 Petite, driven by John Jones. The impact involved primarily the rear of the sedan and the front of the Petite. Mr. Smith first sought medical treatment on the day after the MVA. At that time, his primary complaints were neck and back pain.

DISCUSSION

The medical records indicate that Mr. Smith was 5'10" tall and weighed 205 lb. The MVA was minor and would not be expected to produce significant occupant motion. The absence of any cuts or bruises after the MVA indicated that Mr. Smith had no significant impacts with the vehicle interior, and his x-rays did not show any injuries that would have indicated that injury-producing neck motions had occurred. Both vehicles sustained minor damage; the repair estimate (including labor) was less than $200 for the sedan and less than $500 for the Petite.

A widely used and well-established[1, 2] measure of crash severity is "delta v," a number representing the speed at which a reference crash would produce similar vehicle damage. (If an exemplar vehicle were to impact a solid wall at a speed of delta v, for example, it would replicate the vehicle damage that occurred in the original impact.) Based on vehicle deformation, the damage due to the subject impact was minor and consistent with an impact of less than 5 mph delta v. This is quite comparable to the delta v produced in amusement park bumper car impacts.[3] Note that bumper car occupants are of a wide range of ages and frailty and typically experience repeated impacts, from multiple directions, without injury. Thus, vehicle impacts of severity comparable to this incident are readily withstood by children and the elderly. Also, a series of rear-impact human volunteer tests, involving multiple exposures and testing at speeds greater than 5 mph delta v, produced no complaints other than temporary discomfort that resolved within a few days. [4]

A series of x-rays, taken the day after the MVA, was interpreted by the radiologist as indicating no injury to the cervical (neck), thoracic (upper back), and lumbar/sacral (lower back) regions. The plaintiff then received a second series of x-rays approximately three weeks later. Again, this confirmed that there was no bone or disc injury. The radiology report for the later set of x-rays noted "cervical straightening," that is, the neck was less curved, than what is "normal." It has been claimed in the Beta report (February 3, 2010) that this radiographic finding is relevant to injury analysis, the basic reasoning being that cervical straightening could serve as a marker for trauma-induced muscle spasm. That is, if the motor vehicle impact produced head motion that was sufficiently abrupt to stretch some of the neck muscles until they responded with spasm, the muscle spasm could tighten and straighten the normally curved cervical spine. However, in practice, a straightened neck appearance on x-ray typically reflects the normal variation in neck curvature among the population or is "positional," that is, a temporary, non-traumatic neck straightening due to head positioning at the time of imaging (moving the chin toward the chest tends to straighten the neck)[5–7] and generally is not regarded as a reliable indicator of traumatic injury. Indeed, for such a mechanism to be plausible biomechanically, it would have to be accompanied by clinical correlates (which was not the case in this instance). Furthermore, if muscle tightening altered the spine's curvature, it might tend to increase the lordotic curvature of the spine, rather than straighten it.[8]

In any event, for the subject incident, the earlier x-rays did not detect any straightening; thus, any straightening seen on the later x-rays should be attributed to an event that occurred between the first and second set of x-rays. Of course, this would exclude the subject motor vehicle impact, which occurred before the first set of x-rays was taken. Assuming there was no additional neck trauma between the first and second set of x-rays, biomechanically, any straightening that may have appeared on the later set of x-rays should be considered as positional and therefore benign, transient, and not trauma-related.

Note that the medical records from Mr. Smith's second ER visit indicate that he was in no apparent distress, showed "no evidence of trauma," and his pain was assessed as "minor." Contrary to this, his subjective assessment indicated his pain was extremely severe—10 out of a maximum of 10. This was in marked contrast to the objective notes from the ER staff regarding his pain-related behaviors, which were more consistent with pain assessments in the 2–3 range.

Thank you for this consultation.

REFERENCES

1. Fricke, L.B., *et al.*, *Traffic Accident Investigation Manual*, Northwestern University, Evanston, 1990.

2. Supplemental Notice of Proposed Rulemaking, Docket No. NHTSA 99-6407; Notice 1, FMVSS 208, Occupant Crash Protection, National Highway Traffic Safety Administration, Washington, DC, 1999.

3. Siegmund, G.P., *et al.*, "Speed Change (Δv) of Amusement Park Bumper Cars," *Proceedings of the Canadian Multidisciplinary Road Safety Conference*, Saskatoon, June 14–16, 1993.

4. McConnell, W.E., *et al.*, "Human Head and Neck Kinematics after Low Velocity Rear-End Impacts: Understanding Whiplash," SAE Paper No. 952724, Society of Automotive Engineers, Warrendale, PA, 1995.

5. Harris, J.H., *et al.*, *The Radiology of Acute Cervical Trauma,* 2nd Ed., Williams & Wilkins, Baltimore, 1987.

6. Helliwell, P.S., *et al.*, "The Straight Cervical Spine: Does It Indicate Muscle Spasm," *J. Bone Joint Surg. [Br]*, 76, B:103–106, 1994.

7. Bontrager, K.L., *Textbook of Radiographic Positioning and Related Anatomy*, 3rd Ed., Mosby Year Book, St. Louis, 1993.

8. Zohn, D.A., *Musculoskeletal Pain: Diagnosis and Physical Treatment*, 2nd Ed., Little Brown, Boston, 1988.

Example 5

The basic scenario for Example 5 is that a man fell from a vehicle that was parked in a display area. He landed on his head and received a fatal brain injury. The consultant addressed the biomechanical injury causation question of "How would the subject's fatal head injury have been affected if the surface on which he had landed had been more compliant?" Although at first it may seem intuitively obvious that a softer floor would have been beneficial, in this particular instance, the consultant concluded that a softer floor would not have mitigated the fatal head injury. In essence, the consultant reasoned that the brain injury was hypoxic (due to lack of oxygen) and not due to mechanical head trauma per se. More specifically, the head–floor impact only served to stop the head, and the momentum of the falling torso produced a diving-type neck injury (Pike, 2002) that interfered with respiration and in turn produced hypoxic brain injury. Thus, in this instance, a softer floor would not have mitigated the injury. Note that this report has a slightly different format—specific medical records are identified and cited as references. Although not typical, another possibility would be to include excerpts from the referenced records as attachments, thereby making the report more "self-contained" and making the supporting records more readily accessible.

CONSULTATION REPORT

ABC Consulting, Inc.
1234 Main Street
Anytown, MI 48111
Phone: (313) 123-4567
Fax: (313) 456-7890
Expert1@ABCconsult.com

To:	Ann Attorney, Esq.
E-mail:	aattorney@lawfirm.com
Phone:	(987) 654-3210
Fax:	(987) 654-3211
From:	E. Expert ABC Consulting, Inc.
E-mail:	Expert1@ABCconsult.com
Phone:	(313) 123-4567
Fax:	(313) 456-7890
Date:	June 30, 2009
Subject:	John Doe (deceased)
DOI:	September 31, 2008
YOB:	1973
Height:	5'9"
Weight:	255 lb.

OPINIONS

1. The subject was "prepping" a vehicle (a convertible) for display.
2. While standing on the rear seat, the subject fell over the side of the vehicle and landed head first on the floor.
3. The subject died from a fatal brain injury.
4. The head-ground impact per se did not produce significant head or brain injury.

5. A softer ground contact surface would not have mitigated the fatal injury in this instance.

The opinions are stated to a reasonable degree of biomechanical certainty. My experience and qualifications are outlined in my attached CV. The comments are based on the material listed here and may be revised if additional material becomes available.

MATERIAL

Medical records, including the following:

A-1 Ambulance run sheet

Regional Hospital Emergency Department

Regional Hospital Radiology Department

Academic Radiology Associates/J. Jones, M.D.

Deposition of Able Baker (November 17, 2008)

Deposition of Sigma Delta (November 17, 2008)

Deposition of Jane Smith, M.D. (January 5, 2009)

BACKGROUND

The plaintiff fell from a stationary vehicle and landed on his head. He was unconscious at the scene. He remained unconscious and died six days after the fall.

DISCUSSION

The subject was working on preparing a stationary vehicle (a convertible, with the top retracted and not covering the vehicle). While standing on the rear seat, the subject fell out of the vehicle, striking the top and right side of his head on the floor. The mechanism by which he lost his balance has not been established, and so, the extent to which the fall was caused by internal (medical) or external (environmental) factors is not known. In general, internal factors (e.g., neurosensory deficit, medication side effect) may produce a tumble (gradual balance loss with diminished postural correction ability) or a crumple (e.g., faint or drop attack), and external factors may produce a slip (foot interaction with a low-friction walking surface) or a trip (foot interaction with an obstacle or loose footwear). These may also involve a stumble—one or several steps taken to regain balance.(1–5)

In any event, the head-floor impact did not produce any significant head or brain injury. An initial CT of the head(6) reported no intracranial bleeding, and this was consistent with a second head CT[7] that went on to report a "negative computerized tomography of the head." This was followed by a head MRI that was read as "No significant intracranial abnormality is seen."[8] An MRI of the cervical spine detected anterior wedge fractures of C4, body and lamina fracture of C5, and several pre-existing spinal irregularities, including widespread degenerative changes and an extremely narrow canal.[9]

After his head contacted the ground, the subject's body continued to "fall" and to apply loading to his neck. Thus, the effect of striking the head primarily was to stop the motion of the head rather than to produce head or brain injury per se. The neck loads and movements, in conjunction with the anatomic and physiologic susceptibility to spinal cord injury, compromised the brain's oxygen supply. In turn, this produced hypoxic brain injury.[10] Note that a similar neck injury mechanism is seen in diving accidents, where the head–bottom impact produces little or no head injury, but the subsequent neck loading produces spinal cord injury.[11]

In summary, although the patient fell on his head and died from a brain injury, the head impact per se did not cause the brain injury. Rather, the brain injury resulted from the neck-bending-based spine injury that compromised cardiopulmonary function and hence compromised the oxygen supply to the brain. Thus, a softer ground surface would not have mitigated the brain injury.

Thank you for this consultation.

REFERENCES

1. Baker, S.L., "The Preventability of Falls," in Gray, J.A.M., *Prevention of Disease in the Elderly*, Churchill Livingston, New York, 1985.

2. Cape, R.D.T., "Falls," in Rossman, I., *Clinical Geriatrics, 3rd Edition*, Lippincott, Philadelphia, PA, 1986.

3. Grabiner, M.D., *et al.*, "Modeling Recovery from Stumbles: Preliminary Data on Variable Selection and Classification Efficacy," *Journal of the American Geriatric Society*, Vol. 40, No. 9, 1992, pp. 910–913.

4. Lim, M.R., *et al.*, "Evaluation of the Elderly Patient with an Abnormal Gait," *Journal of the American Academy of Orthopaedic Surgeons*, Vol. 15, No. 2, 2007, pp. 107–117.

5. Wieman, H.M., *et al.*, "Falls," in Calkins, E., *et al.*, *The Practice of Geriatrics*, Saunders, Philadelphia, PA, 1986.

6. Head CT 1, 9/31/08.

7. Head CT 2, 9/31/08.

8. Head MRI, 10/1/08.

9. Neck MRI, 10/2/08.

10. Discharge summary, 10/06/08.

11. Pike, J.A., *Neck Injury: The Use of X-Rays, CTs, and MRIs to Study Crash-Related Injury Mechanisms*, Society of Automotive Engineers, Warrendale, PA, 2002.

Example 6

In this example, a vehicle left the road and impacted several roadside objects, including a utility pole and a tree. The driver sustained an eye injury (orbital fracture), and the airbag did not deploy. The consultant addressed two questions: (1) Should the airbags have deployed, and (2) If the airbags had deployed, how would it have affected the patient's injury?

CONSULTATION REPORT

XYZ Biomechanics, LLC
12345 Main Street
Anytown, MI 48111
Phone: (313) 123-4567
Fax: (313) 123-5678
Expert1@xyz.com

To: A. Lawyer, Esq.
Able, Baker, and Delta, LLP

Fax: (111) 123-4567

From: D. Expert

Date: January 1, 2011

Re: Mary Doe (driver)

DOI: June 31, 2009

YOB: 1950

The opinions are stated to a reasonable degree of biomechanical certainty. My experience and qualifications are outlined in my attached CV. The comments are based on the material listed here and may be revised if additional material becomes available.

MATERIAL

Medical records, including the following:

ABC Hospital/A. Baker, M.D.

I. Omega, M.D.

Midtown Physical Medicine and Rehabilitation

Police accident report

Vehicle records

Copies of photographs of the plaintiff and the plaintiff's vehicle

Notes and photos from a vehicle inspection (November 1, 2010)

Jones J., Accident reconstruction report (October 16, 2010)

OPINIONS

1. The driver's seat belt showed load marks on the D-ring and tongue assembly.
2. The driver was wearing her seat belt at the time of the incident.
3. Airbag system deployment was not warranted in this event.
4. If the airbag had deployed during this event, it may have had no effect on the driver's orbital fracture.

BACKGROUND

The plaintiff, Jane Doe, was the driver of a 2007 sedan. While she was at the wheel, the vehicle left the road and impacted several fixed objects, including essentially frontal impacts with a telephone pole and a tree. No other vehicles were involved. The airbags did not deploy.

DISCUSSION

The vehicle inspection indicated the following: (1) the driver's-side sun visor showed no hair transfer, no deformation, and no blood-like stains; (2) the steering assembly showed no deformation and no blood-like stains; (3) the seat belt showed load marks on the D-ring and tongue assembly; (4) the driver's-side lower instrument panel showed scuff marks; and (5) the vehicle was involved in multiple impacts. In accordance with prior agreement regarding protocol for this vehicle inspection, the restraint control module ("black box") was not accessed or read. Due to highway construction, the impact site was not available for inspection.

In general, a vehicle impact tends to move an occupant toward the point of impact. Thus, in a frontal impact, the occupant is moved toward the front; in a rear impact, the occupant is moved toward the rear. The load marks are quite consistent with abdomen and chest bruising from the seat belt restraining forces. Also, lower instrument panel (IP) scuffing is consistent with occupant forward motion that produced knee-IP impact. (The driver was 6'2" tall, and although she was belted, her knee contact with the IP would not be surprising.)

Airbags are designed to deploy in moderate to severe crashes.[(1)] Although the damage to this vehicle may appear to correspond to the lower end of a moderate impact,[(2)] in this instance the damage was produced by two separate impacts, neither of which was sufficient to require airbag deployment. Also note that while airbag deployment reduces the incidence of orbital fracture,[(3)] it cannot protect against all eye injury.

In this particular instance, the airbag did not deploy, and an orbital fracture occurred. However, biomechanically, these two events may not be causally related to each other. Indeed, the fact that no deformation or contact marks were observed on the steering wheel rim, steering wheel hub, airbag, or sun visor raises the likelihood of occupant contact with a structure from outside the vehicle as the mechanism of eye injury.

In this instance, the effect of airbag deployment on Ms. Doe's orbital fracture would have depended on several factors, including the following: (1) during which frontal impact (i.e., tree or pole) did the injury occur, and (2) which structure contacted her face to produce the fracture. That is, the airbag may have deployed during one vehicle impact, and the eye injury may have been produced during another vehicle impact. Also, the eye impact path may have been outside the physical dimensions of the airbag. In either (or both) of these instances, the deployment of the airbag would likely have had little mitigating effect on the eye injury.

It is noted that on the day of the accident, Ms. Doe had taken her daily dose of a new prescription medication[(4)] that has the frequently occurring side effects of dizziness and/or drowsiness.[(5)] For this reason, recipients are cautioned not to drive while taking this medication.

Thank you for this consultation.

REFERENCES

1. U.S. Department of Transportation, National Highway Traffic Safety Administration, "Air Bag Basics," www.safercar.gov.

2. Jones, J., Accident reconstruction report, October 16, 2010.

3. Duma, S., *et al.*, "Eye Injury and Orbital Fracture Patterns in Frontal Automobile Crashes," SAE Paper No. 2003-01-0511, Society of Automotive Engineers, Warrendale, PA, 2003.

4. Doe, M., Statement to police, June 31, 2009.

5. *Drug Facts and Comparisons*, Wolters Kluwer, St. Louis, MO, 2007.

References

Alyas, F., Turner, M., and Connell, D., "MRI Findings in the Lumbar Spines of Asymptomatic Adolescent Elite Tennis Players, *British Journal of Sports Medicine,* Vol. 41, No. 11, November 2007, pp. 836–841.

American Forensics Association, www.americanforensics.org.

Anderson, L. and D'Alonzo, R., "Fractures of the Odontoid Process of the Axis," *J. Bone Joint Surg.*, Vol. 56A, 1974, p. 1663.

Banks, R., *et al.*, "Assessing Injury Potential in Rear-End Vehicle Collisions," in Yoganandan, N. and Pintar, F., *Frontiers in Whiplash Trauma*, IOS, Amsterdam, 2000.

Baxt, W.G., *Trauma: The First Hour*, Appleton, Norwalk, CT, 1985.

Bledsoe, B.E., Porter, R.S., and Cherry, R.A., *Essentials of Paramedic Care*, Prentice Hall, Upper Saddle River, NJ, 2003.

Boden, S.C., Wiesel, S.W., Laws, E.R., *et al.*, "Metabolic Bone Disease and the Spine," in Boden, S.C., *et al.*, *The Aging Spine: Essentials of Pathophysiology, Diagnosis and Treatment*, Saunders, Philadelphia, 1991.

Bohlman, H.H., "The Neck," in D'Ambrosia, R.D. (ed.), *Musculoskeletal Disorders*, 2nd Ed., Lippincott, Philadelphia, 1986.

Daffner, R.H., *Imaging of Vertebral Trauma*, 2nd Ed., Lippincott-Raven Publishers, Philadelphia, 1993.

Duckworth, T., Lecture Notes on Orthopaedics and Fractures, Blackwell, Oxford, 1984.

Fricke, L.B. *et al.*, *Traffic Accident Investigation Manual*, Northwestern University, Evanston, IL, 1990.

Fry, D.E., in Richardson, J.C., *Trauma: Clinical Care and Pathophysiology*, Year Book, Chicago, 1987.

Gorman, W. and Hodak, J., "Herniated Intervertebral Disc Without Pain," *Journal of the Oklahoma State Medical Association*, Vol. 90, 1997, pp. 185–190.

Harris, J.H., *The Radiology of Acute Cervical Spine Trauma*, Williams & Wilkins, Baltimore, 1989.

Harris, J.H., *et al.*, *The Radiology of Acute Cervical Trauma, 2nd Edition,* Williams & Wilkins, Baltimore, MD, 1987.

Harris, J.H., Harris, W.H., and Novelline, R.A., *The Radiology of Emergency Medicine*, 3rd Ed., Williams & Wilkins, Baltimore, 1993.

Harris, J.H. and Mirvis, S.E., *The Radiology of Acute Cervical Spine Trauma*, 3rd Ed., Williams & Wilkins, Baltimore, 1996.

Helliwell, P.S., *et al.*, "The Straight Cervical Spine: Does It Indicate Muscle Spasm?" *Journal of Bone and Joint Surgery [Br.]*, Vol. 76B, 1994, pp. 103–106.

Hockberger, R.S., "Spinal Injury," in Rosen, P., *et al.*, *Emergency Medicine: Concepts and Clinical Practice*, 2nd Ed., Mosby, St Louis, 1988.

Huelke, D.F., "Anatomy of the Human Cervical Spine and Associated Structures," SAE Paper No. 790130, Society of Automotive Engineers, Warrendale, PA, 1979.

Jobe, C.M., "Gross Anatomy of the Shoulder," in Rockwood, C.A., *et al.*, *The Shoulder*, 2nd Ed., Vol. 1, W.B. Saunders, Philadelphia, 1998.

Mahoney, B.D., "Cervical Injuries," in Tintinalli, J.E., *Emergency Medicine*, 2nd Ed., McGraw-Hill, New York, 1988.

McRae, D., "Asymptomatic Intervertebral Disc Protrusions," *Acta Radiol.*, Vol. 46, 1956, pp. 9–27.

Mullins, R.J., *et al.*, "Respiratory Pathophysiology," in Richardson, J.C., *et al.*, *Trauma: Clinical Care and Pathophysiology*, Year Book, Chicago, 1987.

National Highway Traffic Safety Administration, Federal Motor Vehicle Safety Standard 214, Side Impact Protection, National Highway Traffic Safety Administration, Washington, D.C., 1990.

National Highway Traffic Safety Administration, Supplemental Notice of Proposed Rulemaking, Docket No. NHTSA 99-6407; Notice 1, Federal Motor Vehicle Safety Standard 208, Occupant Crash Protection, National Highway Traffic Safety Administration, Washington, D.C., 1999.

Pike, J.A., *Automotive Safety: Anatomy, Injury, Testing, and Regulation*, Society of Automotive Engineers, Warrendale, PA, 1990.

Pike, J.A., *Neck Injury: The Use of X-Rays, CTs, and MRIs to Study Vehicle-Related Injury Mechanisms*, Society of Automotive Engineers, Warrendale, PA, 2002.

Pike, J.A., "The Quantitative Effect of Age on Injury Outcome," Proceedings of the 12th Experimental Safety Vehicle (ESV) Conference, Gothenburg, Sweden, May 30, 1989, Paper No. 89-1A-W-020, National Highway Traffic Safety Administration, Washington, DC, 1989.

Plum, F. and Posner, J.B., *The Diagnosis of Stupor and Coma*, 3rd Ed., F.A. Davis, Philadelphia, 1982.

Preston, D.C. and Shapiro, B.E., *Electromyography and Neuromuscular Disorders*, Butterworth-Heinemann, Boston, 1998.

Steinbach, L.S., *et al.*, *Shoulder Magnetic Resonance Imaging*, Lippincott, Philadelphia, 1998.

Zohn, D.A., *Musculoskeletal Pain: Diagnosis and Physical Treatment, 2nd Edition*, Little, Brown, Boston, 1988.

Appendix A

א

Tables of Figures

The following tables are provided to enable readers to easily locate figures that appear anywhere in the first ten chapters of this book and are of a particular type (e.g., radiology report), topic (e.g., seat belt usage), body region (e.g., cervical spine), injury type (e.g., fracture), and imaging modality (e.g., MRI). The first table is a list of the abbreviations used in the tables that follow it.

TABLE A-1
ABBREVIATIONS

Abbreviation	Terminology	Category
A	Abdomen	Body Region
B	Bone Scan	Image Type
C	Chest	Body Region
CAT	Computed Axial Tomography	Image Type
CT	Computed Tomography	Image Type
CHI	Closed Head Injury	Injury Type
Cns	Consult	Record Type
Cnt	Contusion	Injury Type
Com	Compression	Injury Type
Death Cert	Death Certificate	
Dis	Dislocation	Injury Type
Dsk	Disc	Injury Type
E	Electrodiagnostic	Image Type
EMS	Emergency Medical Services	Record Type
ER	Emergency Room/ Emergency Department	Record Type
Fx	Fracture	Injury Type
G	Degenerative	Injury Type
H	Head	Body Region
Hn	Herniation	Injury Type
Hx	History	Record Type
I	Illustration	Image Type
ICE	Independent Chiropractic Examination	
IME	Independent Medical Examination	Record Type
IP	Instrument Panel	
Lab	Laboratory Test	Image Type
Lac	Laceration	Injury Type
LE	Lower Extremity	Body Region
M	Magnetic Resonance Imaging	Image Type
N	Neck	Body Region
NCV	Nerve Conduction Velocity	
Nu	Neurologic	Injury Type
P	Photo	Image Type
Pel	Pelvis	Body Region
Pn	Pain	Injury Type
Police	Police Report	Record Type
Radiology Rpt	Radiology Report	Record Type

TABLE A-1 *(cont.)*

Abbreviation	Terminology	Category
S	Spine	Body Region
Sc	Spine–Cervical	Body Region
Sl	Spine–Lumbar	Body Region
Sp	Sprain	Injury Type
St	Spine–Thoracic	Body Region
Str	Straightening	Injury Type
Th	Thorax	Body Region
Tm	Thrombosis	Injury Type
U	Ultrasound	Image Type
UE	Upper Extremity	Body Region
X	X-ray	Image Type

TABLE A-2
FIGURES BY BODY REGION, INJURY TYPE, AND IMAGING MODALITY
(* = Background Section of Book)

Figure				Body Region and Injury				Image	
Ch. No.	Fig. No.	Type	Topic	Reg. 1	Inj. 1	Reg. 2	Inj. 2	I1	I2
*	A	Anatomic		Th				I	
*	B	Anatomic		Th				I	
*	C	Anatomic		Th				I	
*	D	Anatomic		S				I	
*	E	Anatomic		S				I	
*	F	Anatomic		S	Fx	S	Cnt	I	
1	1	ER		Sc					
1	2	EMS	Belt Use						
1	3	ER							
1	4	ER	Cleared	Sc					
1	5	ER	Cleared	Th	Fx	Sc	Fx		
1	6	Death Cert.	Cleared	Sc					
1	7	Death Cert.	Cause	Sc		Hn			
1	8	Autopsy	Summary	Sc	Fx				
1	9	Autopsy	Pending						
1	10	Autopsy	Update	H					
1	11	Death Cert.	Accident						
1	12	ER	Belt Use	H	CHI				
1	13	ER	Cleared	Sc					
2	1	Sketch	Wavelength						
2	2	X		Sc				X	
2	3	Radiology Rpt.		Sc	Fx	Sc	Dis	X	
2	4	Radiology Rpt.		Sc	Fx	Sc	Dis	X	
2	5	Radiology Rpt.		Sc	Fx			X	
2	6	Radiology Rpt.		Sc	Fx			X	
2	7	History		Sc	Nu				
2	8	History		Sc	Fx				
2	9	Radiology Rpt.		Sc	G			X	
2	10	Radiology Rpt.		Sc	Str			X	
2	11	Radiology Rpt.		Sc	Str			X	
2	12	Radiology Rpt.	Deceleration	Sc				M	

TABLE A-2 *(cont.)*

Figure				Body Region and Injury				Image	
Ch. No.	Fig. No.	Type	Topic	Reg. 1	Inj. 1	Reg. 2	Inj. 2	I1	I2
2	13	Radiology Rpt.		Sc	Fx			X	
2	14	Radiology Rpt.		Sc	Fx			X	
2	15	Radiology Rpt.		Sl	Hn	Sl	Fx	M	
2	16	Radiology Rpt.		S				B	
2	17	Radiology Rpt.		S				B	
2	18	Radiology Rpt.		Th	Fx			B	
2	19	Radiology Rpt.		LE	Tm			U	
2	20	Radiology Rpt.		LE	Tm			U	
2	21	Radiology Rpt.		Th				U	
2	22	Radiology Rpt.		UE				M	
3	1	ER	Intoxication	LE	Lac				
3	2	Lab	Intoxication						
3	3	Toxicology	Drugs						
3	4	NCV		Sc	Nu			E	
3	5	NCV		Sc	Nu			E	
3	6	NCV		Sc	Nu			E	
3	7	Neuropsychology		H	Nu			N	
3	8	Cns		H	Nu			N	
3	9	Radiology Rpt.		H	Nu			M	
4	1	ER	Seat Belt	H		Th			
4	2	Cns	Seat Belt	LE	Fx				
4	3	ER	Seat Belt	A	Cnt				
4	4	ER	Seat Belt	A	Cnt				
4	5	V-Lower IP		LE				P	
4	6	V-Lower IP		LE					
4	7	V-Upper IP		H					
4	8	V-Upper IP		H					
4	9	V-Sun Visor		H					
4	10	V-Sun Visor		H					
4	11	V-Windshield		H					
4	12	V-Windshield		H					
4	13	V-Steering Wheel							
4	14	V-Latch Plate 1	Seat Belt						

TABLE A-2 *(cont.)*

Figure				Body Region and Injury				Image	
Ch. No.	Fig. No.	Type	Topic	Reg. 1	Inj. 1	Reg. 2	Inj. 2	I1	I2
4	15	V-Latch Plate 2	Seat Belt						
4	16	V-Latch Plate 3	Seat Belt						
4	17	V-D Ring	Seat Belt						
4	18	V-Belt Webbing	Seat Belt						
4	19	V-Occupant	Seat Belt	Th	Cnt			P	
4	20	V-Airbag	Airbag						
4	21	ER	Seat Belt						
4	22	ER	Seat Belt						
4	23	Cns	Seat Belt						
4	24	ER	Seat Belt	A	Cnt				
4	25	Cns	Seat Belt	Pel	Fx			C	
4	26	Cns	Seat Belt						
4	27	EMS	Seat Belt	H	CHI				
4	28	ER	Seat Belt	H	CHI				
4	29	EMS	Seat Belt						
4	30	Police	Cell Phone						
5	1	Radiology Rpt.	Degenerative	Sc	G			X	
5	2	Cns	Pre-Existing	H	CHI				
5	3	Radiology Rpt.	Degenerative	LE	G			X	
5	4	ER	Intoxication	LE	F				
5	5	Radiology Rpt.	Pre-Existing	LE				X	
5	6	Radiology Rpt.	Degenerative	UE	G			M	
5	7	Cns	Pre-Existing	H	Nu				
5	8	Cns	Pre-Existing	H	Nu			M	E
5	9	Cns	Pre-Existing	UE	Lac				
5	10	Cns	Pre-Existing	H	Cnt				
5	11	Cns	Pre-Existing	N	Pn				
5	12	Cns	Pre-Existing	Th	Fx			X	
5	13	Cns	Pre-Existing	S					
5	14	Cns	Post-Op	Sc	Fx				
5	15	Cns	Pre-Existing	Sc	Fx			X	
5	16	Cns	Pre-Existing	Sc	Fx				
5	17	Cns	Pre-Existing	Sc	Fx				
5	18	ER	Pre-Existing	Th					

TABLE A-2 *(cont.)*

Figure				Body Region and Injury				Image	
Ch. No.	Fig. No.	Type	Topic	Reg. 1	Inj. 1	Reg. 2	Inj. 2	I1	I2
6	1	EMS	Seat Belt	H					
6	2	Police		H					
6	3	Radiology Rpt.		H	Fx			X	
6	4	Radiology Rpt.		H	Fx			C	
6	5	Cns	History	H					
6	6	Cns	Dizzy	H	Nu	Th		E	M
6	7	Cns	Dizzy	H	Nu			E	
6	8	Cns	Dizzy	H	Nu			C	M
6	9	Cns	Dizzy	H	Nu				
6	10	Cns	Dizzy	H	Nu				
6	11	Police	Pedestrian	LE					
6	12	ER	Pedestrian	LE					
6	13	Cns	NCV	Sc	Dsk	UE	Nu	M	E
6	14	Cns	NCV	Sc	Nu			E	
6	15	Cns	NCV	Sc	Nu			E	
6	16	Cns	NCV	Sc	Nu			E	
6	17	Cns	NCV	Sl	Nu	LE	Nu	E	
6	18	Cns	NCV	Sl	Nu	LE	Nu	E	
6	19	Cns	NCV	Sl	Nu			E	
6	20	Police		H					
6	21	Radiology Rpt.	Spasm	Sc	Str			X	
6	22	Radiology Rpt.		H				M	
6	23	Radiology Rpt.		Sl		Sc		M	
6	24	Radiology Rpt.		H				E	
6	25	Cns		H				E	M
7	1	ER	Pain	N		Th			
7	2	ER	Pain	N					
7	3	ER		Sc				X	
7	4	ER		Sc	Fx			X	
7	5	ER		Sc	Fx			X	
7	6	ER		Th	Fx			X	
7	7	Cns	Congenital	Sc				X	
7	8	Cns		St	G			X	
7	9	Cns	Congenital	Sl	G			X	

TABLE A-2 *(cont.)*

Figure				Body Region and Injury				Image	
Ch. No.	Fig. No.	Type	Topic	Reg. 1	Inj. 1	Reg. 2	Inj. 2	I1	I2
7	10	Radiology Rpt.		Th	Fx	Th	G	X	
7	11	Cns		Th	Fx			X	
7	12	Cns		S					
7	13	Cns	ICE						
7	14	Cns	ICE	S					
7	15	Cns	ICE	S	Sp	Th	Fx		
7	16	Cns	ICE	Th	Fx				
7	17	Cns	ICE	Th	Fx			B	
7	18	Radiology Rpt.		Sc				M	
7	19	Film		Sc				M	
7	20	Time Line						I	
7	21	ER	Pain	N		Th			
7	22	ER	Pain	N					
7	23	ER		Sc				X	
7	24	ER		Sc	Fx			X	
7	25	ER		Sc	Fx			X	
7	26	ER		Th	Fx			X	
7	27	Cns	Congenital	Sc				X	
7	28	Cns		St	G			X	
7	29	Cns	Congenital	Sl	Fx			X	
7	30	Radiology Rpt.		Th	Fx	Th	G	X	
7	31	Cns		Th	Fx			X	
7	32	Cns		S					
7	33	Cns	ICE						
7	34	Cns	ICE	S					
7	35	Cns	ICE	S	Sp	Th	Fx		
7	36	Cns	ICE	Th	Fx				
7	37	Cns	ICE	Th	Fx			B	
7	38	Radiology Rpt.		Sc				M	
7	39	Film		Sc				M	
7	40	Time Line						I	
8	1	ER	Seat Belt	H					
8	2	EMS							
8	3	Radiology Rpt.		Sc		Th		X	

TABLE A-2 *(cont.)*

Figure				Body Region and Injury				Image	
Ch. No.	Fig. No.	Type	Topic	Reg. 1	Inj. 1	Reg. 2	Inj. 2	I1	I2
8	4	Police							
8	5	Toxicology	Intoxication					L	
8	6	Toxicology	Intoxication					L	
8	7	Autopsy		H		Sc			
8	8	Autopsy		H		Sc			
8	9	Autopsy		Sc	Fx	Th	Fx		
8	10	Autopsy		Sc	Fx	Th	Fx		
8	11	Autopsy		H					
8	12	Autopsy		H	Cnt				
8	13	Autopsy		Th					
8	14	Autopsy		H		Th			
8	15	Autopsy		Th					
8	16	Autopsy		H	Nu				
8	17	Autopsy	Accident						
8	18	Death Cert.		H					
8	19	ER	Seat Belt	H					
8	20	EMS							
8	21	Radiology Rpt.		Sc		Th		X	
8	22	Police							
8	23	Toxicology	Intoxication					L	
8	24	Toxicology	Intoxication					L	
8	25	Autopsy		H		Sc			
8	26	Autopsy		H		Sc			
8	27	Autopsy		Sc	Fx	Th	Fx		
8	28	Autopsy		Sc	Fx	Th	Fx		
8	29	Autopsy		H					
8	30	Autopsy		H	Cnt				
8	31	Autopsy		Th					
8	32	Autopsy		H		Th			
8	33	Autopsy		Th					
8	34	Autopsy		H	Nu				
8	35	Autopsy	Accident						
8	36	Death Cert.		H					
8	37	Illustration		H				S	
8	38	Illustration		H				M	

TABLE A-2 *(cont.)*

Figure				Body Region and Injury				Image	
Ch. No.	Fig. No.	Type	Topic	Reg. 1	Inj. 1	Reg. 2	Inj. 2	I1	I2
9	1	Summary		Sc					
9	2	History		Sc					
9	3	Radiology Rpt.		Sc	Fx	Th	Fx		
9	4	Radiology Rpt.		Th		Pel		X	
9	5	Radiology Rpt.		Th	Fx			X	
9	6	Film		Sc				X	
9	7	Radiology Rpt.		Sc	Fx			X	
9	8	Cns		H					
9	9	Summary		Sc					
9	10	History		Sc					
9	11	Radiology Rpt.		Sc	Fx	Th	Fx		
9	12	Radiology Rpt.		Th		Pel		X	
9	13	Radiology Rpt.		Th	Fx			X	
9	14	Film		Sc				X	
9	15	Radiology Rpt.		Sc	Fx			X	
9	16	Cns		H					
9	17	Illustration	Spinal Cord					I	

TABLE A-3
FIGURES BY BODY REGION
(* = Background Section of Book)

Figure				Body Region and Injury				Image	
Ch. No.	Fig. No.	Type	Topic	Reg. 1	Inj. 1	Reg. 2	Inj. 2	I1	I2
4	3	ER	Seat Belt	A	Cnt				
4	4	ER	Seat Belt	A	Cnt				
4	24	ER	Seat Belt	A	Cnt				
8	7	Autopsy		H		Sc			
8	8	Autopsy		H		Sc			
8	25	Autopsy		H		Sc			
8	26	Autopsy		H		Sc			
8	14	Autopsy		H		Th			
8	32	Autopsy		H		Th			
8	11	Autopsy		H					
8	18	Death Cert.		H					
8	29	Autopsy		H					
8	36	Death Cert.		H					
8	37	Illustration		H				S	
8	38	Illustration		H				M	
9	8	Cns		H					
9	16	Cns		H					
1	12	ER	Belt Use	H	CHI				
4	27	EMS	Seat Belt	H	CHI				
4	28	ER	Seat Belt	H	CHI				
5	2	Cns	Pre-Existing	H	CHI				
5	10	Cns	Pre-Existing	H	Cnt				
8	12	Autopsy		H	Cnt				
8	30	Autopsy		H	Cnt				
6	3	Radiology Rpt.		H	Fx			X	
6	4	Radiology Rpt.		H	Fx			C	
6	6	Cns	Dizzy	H	Nu	Th		E	M
3	7	Neuropsychology		H	Nu			N	
3	8	Cns		H	Nu			N	
3	9	Radiology Rpt.		H	Nu			M	
5	7	Cns	Pre-Existing	H	Nu				
5	8	Cns	Pre-Existing	H	Nu			M	E
6	7	Cns	Dizzy	H	Nu			E	

TABLE A-3 *(cont.)*

Figure				Body Region and Injury				Image	
Ch. No.	Fig. No.	Type	Topic	Reg. 1	Inj. 1	Reg. 2	Inj. 2	I1	I2
6	8	Cns	Dizzy	H	Nu			C	M
6	9	Cns	Dizzy	H	Nu				
6	10	Cns	Dizzy	H	Nu				
8	16	Autopsy		H	Nu				
8	34	Autopsy		H	Nu				
4	1	ER	Seat Belt	H		Th			
1	10	Autopsy	Update	H					
4	7	V-Upper IP		H					
4	8	V-Upper IP		H					
4	9	V-Sun Visor		H					
4	10	V-Sun Visor		H					
4	11	V-Windshield		H					
4	12	V-Windshield		H					
6	1	EMS	Seat Belt	H					
6	2	Police		H					
6	5	Cns	History	H					
6	20	Police		H					
6	22	Radiology Rpt.		H				M	
6	24	Radiology Rpt.		H				E	
6	25	Cns		H				E	M
8	1	ER	Seat Belt	H					
8	19	ER	Seat Belt	H					
5	4	ER	Intoxication	LE	F				
4	2	Cns	Seat Belt	LE	Fx				
5	3	Radiology Rpt.	Degenerative	LE	G			X	
3	1	ER	Intoxication	LE	Lac				
2	19	Radiology Rpt.		LE	Tm			U	
2	20	Radiology Rpt.		LE	Tm			U	
4	5	V-Lower IP		LE				P	
4	6	V-Lower IP		LE					
5	5	Radiology Rpt.	Pre-Existing	LE				X	
6	11	Police	Pedestrian	LE					
6	12	ER	Pedestrian	LE					
7	2	ER	Pain	N					
7	22	ER	Pain	N					

TABLE A-3 *(cont.)*

Figure				Body Region and Injury				Image	
Ch. No.	Fig. No.	Type	Topic	Reg. 1	Inj. 1	Reg. 2	Inj. 2	I1	I2
5	11	Cns	Pre-Existing	N	Pn				
7	1	ER	Pain	N		Th			
7	21	ER	Pain	N		Th			
4	25	Cns	Seat Belt	Pel	Fx			C	
7	12	Cns		S					
7	14	Cns	ICE	S					
7	32	Cns		S					
7	34	Cns	ICE	S					
*	F	Anatomic		S	Fx	S	Cnt	I	
7	15	Cns	ICE	S	Sp	Th	Fx		
7	35	Cns	ICE	S	Sp	Th	Fx		
*	D	Anatomic		S				I	
*	E	Anatomic		S				I	
2	16	Radiology Rpt.		S				B	
2	17	Radiology Rpt.		S				B	
5	13	Cns	Pre-Existing	S					
8	3	Radiology Rpt.		Sc		Th		X	
8	21	Radiology Rpt.		Sc		Th		X	
7	3	ER		Sc				X	
7	7	Cns	Congenital	Sc				X	
7	18	Radiology Rpt.		Sc				M	
7	19	Film		Sc				M	
7	23	ER		Sc				X	
7	27	Cns	Congenital	Sc				X	
7	38	Radiology Rpt.		Sc				M	
7	39	Film		Sc				M	
9	2	History		Sc					
9	6	Film		Sc				X	
9	10	History		Sc					
9	14	Film		Sc				X	
6	13	Cns	NCV	Sc	Dsk	UE	Nu	M	E
2	3	Radiology Rpt.		Sc	Fx	Sc	Dis	X	
2	4	Radiology Rpt.		Sc	Fx	Sc	Dis	X	
8	9	Autopsy		Sc	Fx	Th	Fx		
8	10	Autopsy		Sc	Fx	Th	Fx		

TABLE A-3 *(cont.)*

Figure		Type	Topic	Body Region and Injury				Image	
Ch. No.	Fig. No.			Reg. 1	Inj. 1	Reg. 2	Inj. 2	I1	I2
8	27	Autopsy		Sc	Fx	Th	Fx		
8	28	Autopsy		Sc	Fx	Th	Fx		
9	3	Radiology Rpt.		Sc	Fx	Th	Fx		
9	11	Radiology Rpt.		Sc	Fx	Th	Fx		
1	8	Autopsy	Summary	Sc	Fx				
2	5	Radiology Rpt.		Sc	Fx			X	
2	6	Radiology Rpt.		Sc	Fx			X	
2	8	History		Sc	Fx				
2	13	Radiology Rpt.		Sc	Fx			X	
2	14	Radiology Rpt.		Sc	Fx			X	
5	14	Cns	Post-Op	Sc	Fx				
5	15	Cns	Pre-Existing	Sc	Fx			X	
5	16	Cns	Pre-Existing	Sc	Fx				
5	17	Cns	Pre-Existing	Sc	Fx				
7	4	ER		Sc	Fx			X	
7	5	ER		Sc	Fx			X	
7	24	ER		Sc	Fx			X	
7	25	ER		Sc	Fx			X	
9	7	Radiology Rpt.		Sc	Fx			X	
9	15	Radiology Rpt.		Sc	Fx			X	
2	9	Radiology Rpt.		Sc	G			X	
5	1	Radiology Rpt.	Degenerative	Sc	G			X	
2	7	History		Sc	Nu				
3	4	NCV		Sc	Nu			E	
3	5	NCV		Sc	Nu			E	
3	6	NCV		Sc	Nu			E	
6	14	Cns	NCV	Sc	Nu			E	
6	15	Cns	NCV	Sc	Nu			E	
6	16	Cns	NCV	Sc	Nu			E	
2	10	Radiology Rpt.		Sc	Str			X	
2	11	Radiology Rpt.		Sc	Str			X	
6	21	Radiology Rpt.	Spasm	Sc	Str			X	
1	7	Death Cert.	Cause	Sc		Hn			
1	1	ER		Sc					
1	4	ER	Cleared	Sc					

TABLE A-3 *(cont.)*

Figure				Body Region and Injury				Image	
Ch. No.	Fig. No.	Type	Topic	Reg. 1	Inj. 1	Reg. 2	Inj. 2	I1	I2
1	6	Death Cert.	Cleared	Sc					
1	13	ER	Cleared	Sc					
2	2	X		Sc				X	
2	12	Radiology Rpt.	Deceleration	Sc				M	
9	1	Summary		Sc					
9	9	Summary		Sc					
7	29	Cns	Congenital	Sl	Fx			X	
7	9	Cns	Congenital	Sl	G			X	
2	15	Radiology Rpt.		Sl	Hn	Sl	Fx	M	
6	17	Cns	NCV	Sl	Nu	LE	Nu	E	
6	18	Cns	NCV	Sl	Nu	LE	Nu	E	
6	19	Cns	NCV	Sl	Nu			E	
6	23	Radiology Rpt.		Sl		Sc		M	
7	8	Cns		St	G			X	
7	28	Cns		St	G			X	
9	4	Radiology Rpt.		Th		Pel		X	
9	12	Radiology Rpt.		Th		Pel		X	
2	21	Radiology Rpt.		Th				U	
8	13	Autopsy		Th					
8	15	Autopsy		Th					
8	31	Autopsy		Th					
8	33	Autopsy		Th					
4	19	V-Occupant	Seat Belt	Th	Cnt			P	
1	5	ER	Cleared	Th	Fx	Sc	Fx		
7	10	Radiology Rpt.		Th	Fx	Th	G	X	
7	30	Radiology Rpt.		Th	Fx	Th	G	X	
2	18	Radiology Rpt.		Th	Fx			B	
5	12	Cns	Pre-Existing	Th	Fx			X	
7	6	ER		Th	Fx			X	
7	11	Cns		Th	Fx			X	
7	16	Cns	ICE	Th	Fx				
7	17	Cns	ICE	Th	Fx			B	
7	26	ER		Th	Fx			X	
7	31	Cns		Th	Fx			X	
7	36	Cns	ICE	Th	Fx				

TABLE A-3 *(cont.)*

Figure				Body Region and Injury				Image	
Ch. No.	Fig. No.	Type	Topic	Reg. 1	Inj. 1	Reg. 2	Inj. 2	I1	I2
7	37	Cns	ICE	Th	Fx			B	
9	5	Radiology Rpt.		Th	Fx			X	
9	13	Radiology Rpt.		Th	Fx			X	
*	A	Anatomic		Th				I	
*	B	Anatomic		Th				I	
*	C	Anatomic		Th				I	
5	18	ER	Pre-Existing	Th					
5	6	Radiology Rpt.	Degenerative	UE	G			M	
5	9	Cns	Pre-Existing	UE	Lac				
2	22	Radiology Rpt.		UE				M	

TABLE A-4
FIGURES BY INJURY TYPE
(* = Background Section of Book)

Figure				Body Region and Injury				Image	
Ch. No.	Fig. No.	Type	Topic	Reg. 1	Inj. 1	Reg. 2	Inj. 2	I1	I2
1	12	ER	Belt Use	H	CHI				
4	27	EMS	Seat Belt	H	CHI				
4	28	ER	Seat Belt	H	CHI				
5	2	Cns	Pre-Existing	H	CHI				
4	3	ER	Seat Belt	A	Cnt				
4	4	ER	Seat Belt	A	Cnt				
4	19	V-Occupant	Seat Belt	Th	Cnt			P	
4	24	ER	Seat Belt	A	Cnt				
5	10	Cns	Pre-Existing	H	Cnt				
8	12	Autopsy		H	Cnt				
8	30	Autopsy		H	Cnt				
6	13	Cns	NCV	Sc	Dsk	UE	Nu	M	E
5	4	ER	Intoxication	LE	F				
*	F	Anatomic		S	Fx	S	Cnt	I	
2	3	Radiology Rpt.		Sc	Fx	Sc	Dis	X	
2	4	Radiology Rpt.		Sc	Fx	Sc	Dis	X	
1	5	ER	Cleared	Th	Fx	Sc	Fx		
8	9	Autopsy		Sc	Fx	Th	Fx		
8	10	Autopsy		Sc	Fx	Th	Fx		
8	27	Autopsy		Sc	Fx	Th	Fx		
8	28	Autopsy		Sc	Fx	Th	Fx		
9	3	Radiology Rpt.		Sc	Fx	Th	Fx		
9	11	Radiology Rpt.		Sc	Fx	Th	Fx		
7	10	Radiology Rpt.		Th	Fx	Th	G	X	
7	30	Radiology Rpt.		Th	Fx	Th	G	X	
1	8	Autopsy	Summary	Sc	Fx				
2	5	Radiology Rpt.		Sc	Fx			X	
2	6	Radiology Rpt.		Sc	Fx			X	
2	8	History		Sc	Fx				
2	13	Radiology Rpt.		Sc	Fx			X	
2	14	Radiology Rpt.		Sc	Fx			X	
2	18	Radiology Rpt.		Th	Fx			B	
4	2	Cns	Seat Belt	LE	Fx				

TABLE A-4 *(cont.)*

Figure				Body Region and Injury				Image	
Ch. No.	Fig. No.	Type	Topic	Reg. 1	Inj. 1	Reg. 2	Inj. 2	I1	I2
4	25	Cns	Seat Belt	Pel	Fx			C	
5	12	Cns	Pre-Existing	Th	Fx			X	
5	14	Cns	Post-Op	Sc	Fx				
5	15	Cns	Pre-Existing	Sc	Fx			X	
5	16	Cns	Pre-Existing	Sc	Fx				
5	17	Cns	Pre-Existing	Sc	Fx				
6	3	Radiology Rpt.		H	Fx			X	
6	4	Radiology Rpt.		H	Fx			C	
7	4	ER		Sc	Fx			X	
7	5	ER		Sc	Fx			X	
7	6	ER		Th	Fx			X	
7	11	Cns		Th	Fx			X	
7	16	Cns	ICE	Th	Fx				
7	17	Cns	ICE	Th	Fx			B	
7	24	ER		Sc	Fx			X	
7	25	ER		Sc	Fx			X	
7	26	ER		Th	Fx			X	
7	29	Cns	Congenital	Sl	Fx			X	
7	31	Cns		Th	Fx			X	
7	36	Cns	ICE	Th	Fx				
7	37	Cns	ICE	Th	Fx			B	
9	5	Radiology Rpt.		Th	Fx			X	
9	7	Radiology Rpt.		Sc	Fx			X	
9	13	Radiology Rpt.		Th	Fx			X	
9	15	Radiology Rpt.		Sc	Fx			X	
2	9	Radiology Rpt.		Sc	G			X	
5	1	Radiology Rpt.	Degenerative	Sc	G			X	
5	3	Radiology Rpt.	Degenerative	LE	G			X	
5	6	Radiology Rpt.	Degenerative	UE	G			M	
7	8	Cns		St	G			X	
7	9	Cns	Congenital	Sl	G			X	
7	28	Cns		St	G			X	
2	15	Radiology Rpt.		Sl	Hn	Sl	Fx	M	
3	1	ER	Intoxication	LE	Lac				
5	9	Cns	Pre-Existing	UE	Lac				

TABLE A-4 *(cont.)*

Figure				Body Region and Injury				Image	
Ch. No.	Fig. No.	Type	Topic	Reg. 1	Inj. 1	Reg. 2	Inj. 2	I1	I2
6	17	Cns	NCV	Sl	Nu	LE	Nu	E	
6	18	Cns	NCV	Sl	Nu	LE	Nu	E	
2	7	History		Sc	Nu				
3	4	NCV		Sc	Nu			E	
3	5	NCV		Sc	Nu			E	
3	6	NCV		Sc	Nu			E	
3	7	Neuropsychology		H	Nu			N	
3	8	Cns		H	Nu			N	
3	9	Radiology Rpt.		H	Nu			M	
5	7	Cns	Pre-Existing	H	Nu				
5	8	Cns	Pre-Existing	H	Nu			M	E
6	6	Cns	Dizzy	H	Nu	Th		E	M
6	7	Cns	Dizzy	H	Nu			E	
6	8	Cns	Dizzy	H	Nu			C	M
6	9	Cns	Dizzy	H	Nu				
6	10	Cns	Dizzy	H	Nu				
6	14	Cns	NCV	Sc	Nu			E	
6	15	Cns	NCV	Sc	Nu			E	
6	16	Cns	NCV	Sc	Nu			E	
6	19	Cns	NCV	Sl	Nu			E	
8	16	Autopsy		H	Nu				
8	34	Autopsy		H	Nu				
5	11	Cns	Pre-Existing	N	Pn				
7	15	Cns	ICE	S	Sp	Th	Fx		
7	35	Cns	ICE	S	Sp	Th	Fx		
2	10	Radiology Rpt.		Sc	Str			X	
2	11	Radiology Rpt.		Sc	Str			X	
6	21	Radiology Rpt.	Spasm	Sc	Str			X	
2	19	Radiology Rpt.		LE	Tm			U	
2	20	Radiology Rpt.		LE	Tm			U	

TABLE A-5
FIGURES BY IMAGING MODALITY
(* = Background Section of Book)

Figure				Body Region and Injury				Image	
Ch. No.	Fig. No.	Type	Topic	Reg. 1	Inj. 1	Reg. 2	Inj. 2	I1	I2
7	17	Cns	ICE	Th	Fx			B	
7	37	Cns	ICE	Th	Fx			B	
2	16	Radiology Rpt.		S				B	
2	17	Radiology Rpt.		S				B	
2	18	Radiology Rpt.		Th	Fx			B	
6	8	Cns	Dizzy	H	Nu			C	M
4	25	Cns	Seat Belt	Pel	Fx			C	
6	4	Radiology Rpt.		H	Fx			C	
6	6	Cns	Dizzy	H	Nu	Th		E	M
6	25	Cns		H				E	M
3	4	NCV		Sc	Nu			E	
3	5	NCV		Sc	Nu			E	
3	6	NCV		Sc	Nu			E	
6	7	Cns	Dizzy	H	Nu			E	
6	14	Cns	NCV	Sc	Nu			E	
6	15	Cns	NCV	Sc	Nu			E	
6	16	Cns	NCV	Sc	Nu			E	
6	17	Cns	NCV	Sl	Nu	LE	Nu	E	
6	18	Cns	NCV	Sl	Nu	LE	Nu	E	
6	19	Cns	NCV	Sl	Nu			E	
6	24	Radiology Rpt.		H				E	
*	A	Anatomic		Th				I	
*	B	Anatomic		Th				I	
*	C	Anatomic		Th				I	
*	D	Anatomic		S				I	
*	E	Anatomic		S				I	
*	F	Anatomic		S	Fx	S	Cnt	I	
7	20	Time Line						I	
7	40	Time Line						I	
9	17	Illustration	Spinal Cord					I	
8	5	Toxicology	Intoxication					L	
8	6	Toxicology	Intoxication					L	
8	23	Toxicology	Intoxication					L	

TABLE A-5 *(cont.)*

Figure				Body Region and Injury				Image	
Ch. No.	Fig. No.	Type	Topic	Reg. 1	Inj. 1	Reg. 2	Inj. 2	I1	I2
8	24	Toxicology	Intoxication					L	
5	8	Cns	Pre-Existing	H	Nu			M	E
6	13	Cns	NCV	Sc	Dsk	UE	Nu	M	E
2	22	Radiology Rpt.		UE				M	
3	9	Radiology Rpt.		H	Nu			M	
5	6	Radiology Rpt.	Degenerative	UE	G			M	
6	22	Radiology Rpt.		H				M	
6	23	Radiology Rpt.		Sl		Sc		M	
7	18	Radiology Rpt.		Sc				M	
7	19	Film		Sc				M	
7	38	Radiology Rpt.		Sc				M	
7	39	Film		Sc				M	
8	38	Illustration		H				M	
2	12	Radiology Rpt.	Deceleration	Sc				M	
2	15	Radiology Rpt.		Sl	Hn	Sl	Fx	M	
3	7	Neuropsychology		H	Nu			N	
3	8	Cns		H	Nu			N	
4	5	V-Lower IP		LE				P	
4	19	V-Occupant	Seat Belt	Th	Cnt			P	
8	37	Illustration		H				S	
2	20	Radiology Rpt.		LE	Tm			U	
2	21	Radiology Rpt.		Th				U	
2	19	Radiology Rpt.		LE	Tm			U	
2	2	X		Sc				X	
2	3	Radiology Rpt.		Sc	Fx	Sc	Dis	X	
2	4	Radiology Rpt.		Sc	Fx	Sc	Dis	X	
2	5	Radiology Rpt.		Sc	Fx			X	
2	6	Radiology Rpt.		Sc	Fx			X	
2	9	Radiology Rpt.		Sc	G			X	
2	10	Radiology Rpt.		Sc	Str			X	
2	11	Radiology Rpt.		Sc	Str			X	
2	13	Radiology Rpt.		Sc	Fx			X	
2	14	Radiology Rpt.		Sc	Fx			X	
5	1	Radiology Rpt.	Degenerative	Sc	G			X	
5	3	Radiology Rpt.	Degenerative	LE	G			X	

TABLE A-5 ***(cont.)***

Figure				Body Region and Injury				Image	
Ch. No.	Fig. No.	Type	Topic	Reg. 1	Inj. 1	Reg. 2	Inj. 2	I1	I2
5	5	Radiology Rpt.	Pre-Existing	LE				X	
5	12	Cns	Pre-Existing	Th	Fx			X	
5	15	Cns	Pre-Existing	Sc	Fx			X	
6	3	Radiology Rpt.		H	Fx			X	
6	21	Radiology Rpt.	Spasm	Sc	Str			X	
7	3	ER		Sc				X	
7	4	ER		Sc	Fx			X	
7	5	ER		Sc	Fx			X	
7	6	ER		Th	Fx			X	
7	7	Cns	Congenital	Sc				X	
7	8	Cns		St	G			X	
7	9	Cns	Congenital	Sl	G			X	
7	10	Radiology Rpt.		Th	Fx	Th	G	X	
7	11	Cns		Th	Fx			X	
7	23	ER		Sc				X	
7	24	ER		Sc	Fx			X	
7	25	ER		Sc	Fx			X	
7	26	ER		Th	Fx			X	
7	27	Cns	Congenital	Sc				X	
7	28	Cns		St	G			X	
7	29	Cns	Congenital	Sl	Fx			X	
7	30	Radiology Rpt.		Th	Fx	Th	G	X	
7	31	Cns		Th	Fx			X	
8	3	Radiology Rpt.		Sc		Th		X	
8	21	Radiology Rpt.		Sc		Th		X	
9	4	Radiology Rpt.		Th		Pel		X	
9	5	Radiology Rpt.		Th	Fx			X	
9	6	Film		Sc				X	
9	7	Radiology Rpt.		Sc	Fx			X	
9	12	Radiology Rpt.		Th		Pel		X	
9	13	Radiology Rpt.		Th	Fx			X	
9	14	Film		Sc				X	
9	15	Radiology Rpt		Sc	Fx			X	

TABLE A-6
FIGURES BY RECORD TYPE
(* = Background Section of Book)

Figure				Body Region and Injury				Image	
Ch. No.	Fig. No.	Type	Topic	Reg. 1	Inj. 1	Reg. 2	Inj. 2	I1	I2
*	A	Anatomic		Th				I	
*	B	Anatomic		Th				I	
*	C	Anatomic		Th				I	
*	D	Anatomic		S				I	
*	E	Anatomic		S				I	
*	F	Anatomic		S	Fx	S	Cnt	I	
1	8	Autopsy	Summary	Sc	Fx				
1	9	Autopsy	Pending						
1	10	Autopsy	Update	H					
8	7	Autopsy		H		Sc			
8	8	Autopsy		H		Sc			
8	9	Autopsy		Sc	Fx	Th	Fx		
8	10	Autopsy		Sc	Fx	Th	Fx		
8	11	Autopsy		H					
8	12	Autopsy		H	Cnt				
8	13	Autopsy		Th					
8	14	Autopsy		H		Th			
8	15	Autopsy		Th					
8	16	Autopsy		H	Nu				
8	17	Autopsy	Accident						
8	25	Autopsy		H		Sc			
8	26	Autopsy		H		Sc			
8	27	Autopsy		Sc	Fx	Th	Fx		
8	28	Autopsy		Sc	Fx	Th	Fx		
8	29	Autopsy		H					
8	30	Autopsy		H	Cnt				
8	31	Autopsy		Th					
8	32	Autopsy		H		Th			
8	33	Autopsy		Th					
8	34	Autopsy		H	Nu				
8	35	Autopsy	Accident						
3	8	Cns		H	Nu			N	
4	2	Cns	Seat Belt	LE	Fx				

TABLE A-6 *(cont.)*

Figure				Body Region and Injury				Image	
Ch. No.	Fig. No.	Type	Topic	Reg. 1	Inj. 1	Reg. 2	Inj. 2	I1	I2
4	23	Cns	Seat Belt						
4	25	Cns	Seat Belt	Pel	Fx			C	
4	26	Cns	Seat Belt						
5	2	Cns	Pre-Existing	H	CHI				
5	7	Cns	Pre-Existing	H	Nu				
5	8	Cns	Pre-Existing	H	Nu			M	E
5	9	Cns	Pre-Existing	UE	Lac				
5	10	Cns	Pre-Existing	H	Cnt				
5	11	Cns	Pre-Existing	N	Pn				
5	12	Cns	Pre-Existing	Th	Fx			X	
5	13	Cns	Pre-Existing	S					
5	14	Cns	Post-Op	Sc	Fx				
5	15	Cns	Pre-Existing	Sc	Fx			X	
5	16	Cns	Pre-Existing	Sc	Fx				
5	17	Cns	Pre-Existing	Sc	Fx				
6	5	Cns	History	H					
6	6	Cns	Dizzy	H	Nu	Th		E	M
6	7	Cns	Dizzy	H	Nu			E	
6	8	Cns	Dizzy	H	Nu			C	M
6	9	Cns	Dizzy	H	Nu				
6	10	Cns	Dizzy	H	Nu				
6	13	Cns	NCV	Sc	Dsk	UE	Nu	M	E
6	14	Cns	NCV	Sc	Nu			E	
6	15	Cns	NCV	Sc	Nu			E	
6	16	Cns	NCV	Sc	Nu			E	
6	17	Cns	NCV	Sl	Nu	LE	Nu	E	
6	18	Cns	NCV	Sl	Nu	LE	Nu	E	
6	19	Cns	NCV	Sl	Nu			E	
6	25	Cns		H				E	M
7	7	Cns	Congenital	Sc				X	
7	8	Cns		St	G			X	
7	9	Cns	Congenital	Sl	G			X	
7	11	Cns		Th	Fx			X	
7	12	Cns		S					
7	13	Cns	ICE						

TABLE A-6 *(cont.)*

Figure				Body Region and Injury				Image	
Ch. No.	Fig. No.	Type	Topic	Reg. 1	Inj. 1	Reg. 2	Inj. 2	I1	I2
7	14	Cns	ICE	S					
7	15	Cns	ICE	S	Sp	Th	Fx		
7	16	Cns	ICE	Th	Fx				
7	17	Cns	ICE	Th	Fx			B	
7	27	Cns	Congenital	Sc				X	
7	28	Cns		St	G			X	
7	29	Cns	Congenital	Sl	Fx			X	
7	31	Cns		Th	Fx			X	
7	32	Cns		S					
7	33	Cns	ICE						
7	34	Cns	ICE	S					
7	35	Cns	ICE	S	Sp	Th	Fx		
7	36	Cns	ICE	Th	Fx				
7	37	Cns	ICE	Th	Fx			B	
9	8	Cns		H					
9	16	Cns		H					
8	18	Death Cert.		H					
8	36	Death Cert.		H					
1	6	Death Cert.	Cleared	Sc					
1	7	Death Cert.	Cause	Sc		Hn			
1	11	Death Cert.	Accident						
1	2	EMS	Belt Use						
4	27	EMS	Seat Belt	H	CHI				
4	29	EMS	Seat Belt						
6	1	EMS	Seat Belt	H					
8	2	EMS							
8	20	EMS							
4	4	ER	Seat Belt	A	Cnt				
1	1	ER		Sc					
1	3	ER							
1	4	ER	Cleared	Sc					
1	5	ER	Cleared	Th	Fx	Sc	Fx		
1	12	ER	Belt Use	H	CHI				
1	13	ER	Cleared	Sc					
3	1	ER	Intoxication	LE	Lac				

TABLE A-6 *(cont.)*

Figure				Body Region and Injury				Image	
Ch. No.	Fig. No.	Type	Topic	Reg. 1	Inj. 1	Reg. 2	Inj. 2	I1	I2
4	1	ER	Seat Belt	H		Th			
4	3	ER	Seat Belt	A	Cnt				
4	21	ER	Seat Belt						
4	22	ER	Seat Belt						
4	24	ER	Seat Belt	A	Cnt				
4	28	ER	Seat Belt	H	CHI				
5	4	ER	Intoxication	LE	F				
5	18	ER	Pre-Existing	Th					
6	12	ER	Pedestrian	LE					
7	1	ER	Pain	N		Th			
7	2	ER	Pain	N					
7	3	ER		Sc				X	
7	4	ER		Sc	Fx			X	
7	5	ER		Sc	Fx			X	
7	6	ER		Th	Fx			X	
7	21	ER	Pain	N		Th			
7	22	ER	Pain	N					
7	23	ER		Sc				X	
7	24	ER		Sc	Fx			X	
7	25	ER		Sc	Fx			X	
7	26	ER		Th	Fx			X	
8	1	ER	Seat Belt	H					
8	19	ER	Seat Belt	H					
7	19	Film		Sc				M	
7	39	Film		Sc				M	
9	6	Film		Sc				X	
9	14	Film		Sc				X	
2	7	History		Sc	Nu				
2	8	History		Sc	Fx				
9	2	History		Sc					
9	10	History		Sc					
8	37	Illustration		H				S	
8	38	Illustration		H				M	
9	17	Illustration	Spinal Cord					I	
3	2	Lab	Intoxication						

TABLE A-6 *(cont.)*

Figure				Body Region and Injury				Image	
Ch. No.	Fig. No.	Type	Topic	Reg. 1	Inj. 1	Reg. 2	Inj. 2	I1	I2
3	4	NCV		Sc	Nu			E	
3	5	NCV		Sc	Nu			E	
3	6	NCV		Sc	Nu			E	
3	7	Neuropsychology		H	Nu			N	
4	30	Police	Cell Phone						
6	2	Police		H					
6	11	Police	Pedestrian	LE					
6	20	Police		H					
8	4	Police							
8	22	Police							
2	3	Radiology Rpt.		Sc	Fx	Sc	Dis	X	
2	4	Radiology Rpt.		Sc	Fx	Sc	Dis	X	
2	5	Radiology Rpt.		Sc	Fx			X	
2	6	Radiology Rpt.		Sc	Fx			X	
2	9	Radiology Rpt.		Sc	G			X	
2	10	Radiology Rpt.		Sc	Str			X	
2	11	Radiology Rpt.		Sc	Str			X	
2	12	Radiology Rpt.	Deceleration	Sc				M	
2	13	Radiology Rpt.		Sc	Fx			X	
2	14	Radiology Rpt.		Sc	Fx			X	
2	15	Radiology Rpt.		Sl	Hn	Sl	Fx	M	
2	16	Radiology Rpt.		S				B	
2	17	Radiology Rpt.		S				B	
2	18	Radiology Rpt.		Th	Fx			B	
2	19	Radiology Rpt.		LE	Tm			U	
2	20	Radiology Rpt.		LE	Tm			U	
2	21	Radiology Rpt.		Th				U	
2	22	Radiology Rpt.		UE				M	
3	9	Radiology Rpt.		H	Nu			M	
5	1	Radiology Rpt.	Degenerative	Sc	G			X	
5	3	Radiology Rpt.	Degenerative	LE	G			X	
5	5	Radiology Rpt.	Pre-Existing	LE				X	
5	6	Radiology Rpt.	Degenerative	UE	G			M	
6	3	Radiology Rpt.		H	Fx			X	
6	4	Radiology Rpt.		H	Fx			C	

TABLE A-6 *(cont.)*

Figure				Body Region and Injury				Image	
Ch. No.	Fig. No.	Type	Topic	Reg. 1	Inj. 1	Reg. 2	Inj. 2	I1	I2
6	21	Radiology Rpt.	Spasm	Sc	Str			X	
6	22	Radiology Rpt.		H				M	
6	23	Radiology Rpt.		Sl		Sc		M	
6	24	Radiology Rpt.		H				E	
7	10	Radiology Rpt.		Th	Fx	Th	G	X	
7	18	Radiology Rpt.		Sc				M	
7	30	Radiology Rpt.		Th	Fx	Th	G	X	
7	38	Radiology Rpt.		Sc				M	
8	3	Radiology Rpt.		Sc		Th		X	
8	21	Radiology Rpt.		Sc		Th		X	
9	3	Radiology Rpt.		Sc	Fx	Th	Fx		
9	4	Radiology Rpt.		Th		Pel		X	
9	5	Radiology Rpt.		Th	Fx			X	
9	7	Radiology Rpt.		Sc	Fx			X	
9	11	Radiology Rpt.		Sc	Fx	Th	Fx		
9	12	Radiology Rpt.		Th		Pel		X	
9	13	Radiology Rpt.		Th	Fx			X	
9	15	Radiology Rpt.		Sc	Fx			X	
2	1	Sketch	Wavelength						
9	1	Summary		Sc					
9	9	Summary		Sc					
7	20	Time Line						I	
7	40	Time Line						I	
3	3	Toxicology	Drugs						
8	5	Toxicology	Intoxication					L	
8	6	Toxicology	Intoxication					L	
8	23	Toxicology	Intoxication					L	
8	24	Toxicology	Intoxication					L	
4	20	V-Airbag	Airbag						
4	18	V-Belt Webbing	Seat Belt						
4	17	V-D Ring	Seat Belt						
4	14	V-Latch Plate 1	Seat Belt						
4	15	V-Latch Plate 2	Seat Belt						
4	16	V-Latch Plate 3	Seat Belt						
4	5	V-Lower IP		LE				P	

TABLE A-6 *(cont.)*

Figure				Body Region and Injury				Image	
Ch. No.	**Fig. No.**	**Type**	**Topic**	**Reg. 1**	**Inj. 1**	**Reg. 2**	**Inj. 2**	**I1**	**I2**
4	6	V-Lower IP		LE					
4	19	V-Occupant	Seat Belt	Th	Cnt			P	
4	13	V-Steering Wheel							
4	9	V-Sun Visor		H					
4	10	V-Sun Visor		H					
4	7	V-Upper IP		H					
4	8	V-Upper IP		H					
4	11	V-Windshield		H					
4	12	V-Windshield		H					
2	2	X		Sc				X	

TABLE A-7
FIGURES BY TOPIC
(* = Background Section of Book)

Figure				Body Region and Injury				Image	
Ch. No.	Fig. No.	Type	Topic	Reg. 1	Inj. 1	Reg. 2	Inj. 2	I1	I2
1	11	Death Cert.	Accident						
8	17	Autopsy	Accident						
8	35	Autopsy	Accident						
4	20	V-Airbag	Airbag						
1	2	EMS	Belt Use						
1	12	ER	Belt Use	H	CHI				
1	7	Death Cert.	Cause	Sc		Hn			
4	30	Police	Cell Phone						
1	4	ER	Cleared	Sc					
1	5	ER	Cleared	Th	Fx	Sc	Fx		
1	6	Death Cert.	Cleared	Sc					
1	13	ER	Cleared	Sc					
7	7	Cns	Congenital	Sc				X	
7	9	Cns	Congenital	Sl	G			X	
7	27	Cns	Congenital	Sc				X	
7	29	Cns	Congenital	Sl	Fx			X	
2	12	Radiology Rpt.	Deceleration	Sc				M	
5	1	Radiology Rpt.	Degenerative	Sc	G			X	
5	3	Radiology Rpt.	Degenerative	LE	G			X	
5	6	Radiology Rpt.	Degenerative	UE	G			M	
6	6	Cns	Dizzy	H	Nu	Th		E	M
6	7	Cns	Dizzy	H	Nu			E	
6	8	Cns	Dizzy	H	Nu			C	M
6	9	Cns	Dizzy	H	Nu				
6	10	Cns	Dizzy	H	Nu				
3	3	Toxicology	Drugs						
6	5	Cns	History	H					
7	13	Cns	ICE						
7	14	Cns	ICE	S					
7	15	Cns	ICE	S	Sp	Th	Fx		
7	16	Cns	ICE	Th	Fx				
7	17	Cns	ICE	Th	Fx			B	
7	33	Cns	ICE						

TABLE A-7 *(cont.)*

Figure				Body Region and Injury				Image	
Ch. No.	Fig. No.	Type	Topic	Reg. 1	Inj. 1	Reg. 2	Inj. 2	I1	I2
7	34	Cns	ICE	S					
7	35	Cns	ICE	S	Sp	Th	Fx		
7	36	Cns	ICE	Th	Fx				
7	37	Cns	ICE	Th	Fx			B	
3	1	ER	Intoxication	LE	Lac				
3	2	Lab	Intoxication						
5	4	ER	Intoxication	LE	F				
8	5	Toxicology	Intoxication					L	
8	6	Toxicology	Intoxication					L	
8	23	Toxicology	Intoxication					L	
8	24	Toxicology	Intoxication					L	
6	13	Cns	NCV	Sc	Dsk	UE	Nu	M	E
6	14	Cns	NCV	Sc	Nu			E	
6	15	Cns	NCV	Sc	Nu			E	
6	16	Cns	NCV	Sc	Nu			E	
6	17	Cns	NCV	Sl	Nu	LE	Nu	E	
6	18	Cns	NCV	Sl	Nu	LE	Nu	E	
6	19	Cns	NCV	Sl	Nu			E	
7	1	ER	Pain	N		Th			
7	2	ER	Pain	N					
7	21	ER	Pain	N		Th			
7	22	ER	Pain	N					
6	11	Police	Pedestrian	LE					
6	12	ER	Pedestrian	LE					
1	9	Autopsy	Pending						
5	14	Cns	Post-Op	Sc	Fx				
5	2	Cns	Pre-Existing	H	CHI				
5	5	Radiology Rpt.	Pre-Existing	LE				X	
5	7	Cns	Pre-Existing	H	Nu				
5	8	Cns	Pre-Existing	H	Nu			M	E
5	9	Cns	Pre-Existing	UE	Lac				
5	10	Cns	Pre-Existing	H	Cnt				
5	11	Cns	Pre-Existing	N	Pn				
5	12	Cns	Pre-Existing	Th	Fx			X	
5	13	Cns	Pre-Existing	S					

TABLE A-7 ***(cont.)***

Figure				Body Region and Injury				Image	
Ch. No.	Fig. No.	Type	Topic	Reg. 1	Inj. 1	Reg. 2	Inj. 2	I1	I2
5	15	Cns	Pre-Existing	Sc	Fx			X	
5	16	Cns	Pre-Existing	Sc	Fx				
5	17	Cns	Pre-Existing	Sc	Fx				
5	18	ER	Pre-Existing	Th					
4	14	V-Latch Plate 1	Seat Belt						
4	15	V-Latch Plate 2	Seat Belt						
4	16	V-Latch Plate 3	Seat Belt						
4	17	V-D Ring	Seat Belt						
4	18	V-Belt Webbing	Seat Belt						
4	19	V-Occupant	Seat Belt	Th	Cnt			P	
6	1	EMS	Seat Belt	H					
4	1	ER	Seat Belt	H		Th			
4	2	Cns	Seat Belt	LE	Fx				
4	3	ER	Seat Belt	A	Cnt				
4	4	ER	Seat Belt	A	Cnt				
4	21	ER	Seat Belt						
4	22	ER	Seat Belt						
4	23	Cns	Seat Belt						
4	24	ER	Seat Belt	A	Cnt				
4	25	Cns	Seat Belt	Pel	Fx			C	
4	26	Cns	Seat Belt						
4	27	EMS	Seat Belt	H	CHI				
4	28	ER	Seat Belt	H	CHI				
4	29	EMS	Seat Belt						
8	1	ER	Seat Belt	H					
8	19	ER	Seat Belt	H					
6	21	Radiology Rpt.	Spasm	Sc	Str			X	
9	17	Illustration	Spinal Cord					I	
1	8	Autopsy	Summary	Sc	Fx				
1	10	Autopsy	Update	H					
2	1	Sketch	Wavelength						

Appendix B

א

Glossary

$\bar{A}$	before
AC	acromioclavicular joint
ADL	activities of daily living
AIS	abbreviated injury scale—ordinal scale, integers from 0 (no injury) to 6 (most severe)
AMA	against medical advice
A&Ox3	alert and oriented times three (oriented to person, place, and time)
AP	anterior-posterior (oriented front to back)
AVM	arterio-venous malformation
BAC	blood alcohol concentration
BP	blood pressure
BS	bowel sounds
$\bar{C}$	with
CA	cancer
CAD	coronary artery disease

CAT	computed axial tomography (*also* CT)
C/C	chief complaint
CCX	coccygeal vertebrae (*also* Co)
CHART	chief complaint, history, assessment, Rx, transport
CHF	congestive heart failure
CNS	central nervous system
C/O	complains of
Co	coccygeal vertebrae (*also* CCX)
COPD	chronic obstructive pulmonary disease
CPR	cardiopulmonary resuscitation
C/S	cervical spine (*also* C-spine)
CSF	cerebrospinal fluid
C-spine	cervical spine (*also* C/S)
CT	computed tomography (*also* CAT)
CTS	carpal tunnel syndrome
CV	cardiovascular
CVA	cerebrovascular accident
Cl	first cervical vertebra (*also* C-1)
C-l	first cervical vertebra (*also* C1)
D/C	discharged
DM	diabetes mellitus
DOB	date of birth
DVT	deep vein thrombosis
Dx	diagnosis
ECG	electrocardiogram

ED	emergency department
EDR	electronic data recorder
EEG	electroencephalogram
EENT	eyes, ears, nose, and throat
EKG	electrocardiogram
EMG	electromyeogram
EMS	emergency medical services
ENG	electronystagmogram
ENT	ear, nose, and throat specialist
ER	emergency room
EtOH	ethyl alcohol
Eval	evaluation
F/H	family history (*also* FH)
FH	family history (*also* F/H)
FUO	fever, unknown origin
GCS	Glasgow coma scale—ordinal scale, integers ranging from 3 (most severe) to 15 (least severe); *also* Glasgow coma score
GI	gastrointestinal
HA	headache (*also* Ha)
Ha	headache (*also* HA)
Hb	hemoglobin (*also* Hgb)
HEENT	head, ears, eyes, nose, and throat
Hgb	hemoglobin (*also* Hb)
HNP	herniated nucleus pulposis (ruptured disc)

HTN	hypertension
Hx	history
ICE	independent chiropractic examination
ICU	intensive care unit
IME	independent medical examination
ISS	injury severity score
IV	intravenous
IVC	inferior vena cava
KUB	kidney, ureter, bladder
Ⓛ	left
LBP	low back pain
LE	lower extremity
LLE	left lower extremity
LOC	loss of consciousness
L/S	lumbar spine (*also* L-spine)
L-spine	lumbar spine (*also* L/S)
Lt	left
LUE	left upper extremity
L1	first lumbar vertebra
MAE	moves all extremities
MI	myocardial infarction
MR	magnetic resonance imaging (*also* MRI)

MRI	magnetic resonance imaging (*also* MR)
MVA	motor vehicle accident
MVC	motor vehicle crash (some consider this to be a more neutral designation than MVA)
NAD	no apparent distress
NARD	no apparent respiratory distress
NCR	nerve conduction report
NCS	nerve conduction study
NCV	nerve conduction velocity
ND	non-distended
NIIDM	non-insulin-dependent diabetes mellitus
NKDA	no known drug allergies
Nl	normal
NSR	normal sinus rhythm
NT	non-tender
O	objective
OBGYN	obstetrics/gynecology
OP	outpatient
OR	operating room; *also* open reduction
ORIF	open reduction internal fixation
OT	occupational therapist
OW	otherwise
$\overline{\text{P}}$	after (post)

PA	physician's assistant (may appear as "P.A."; may be followed by a hyphen and a letter a, b, or c)
Palp	palpation
PAR	police accident report
Para	paraplegia
PCR	pre-hospital care report
PE	physical examination
PERL	pupils equal, round, reactive to light and accommodation (*also* pupils equal and reactive to light; PERRLA)
PERRLA	pupils equal, round, reactive to light and accommodation (*also* PERL)
PH	past history (may appear as "P.H.")
PMH	past medical history
PM&R	physical medicine and rehabilitation
PRN	as needed
Pt	patient
PT	physical therapy
qd	every day
qh	every hour
qid	four times per day
Ⓡ	right
RA	rheumatoid arthritis
RBC	red blood cell count
RLE	right lower extremity
RN	registered nurse
R/O	rule out

ROM	range of motion
RRR	regular rate and rhythm
RT	respiratory therapist
Rt	right
RUE	right upper extremity
$\overline{S}$	without (sans)
SLAP	superior labrum anterior-posterior
SOAP	subjective objective assessment plan
SOB	shortness of breath
S/P	status post
Sx	symptoms
TAR	traffic accident report
TENS	transcutaneous electrical nerve stimulator
Th-spine	thoracic spine
TIA	transient ischemic attack
TMJ	temporomandibular joint
Tx	treatment
T1	first thoracic vertebra
UE	upper extremity
US	ultrasound
UTI	urinary tract infection
UV	ultraviolet

VS	vital signs
WBC	white blood cell count
WNL	within normal limits
X2	times two
X4	times four (e.g., MAEX4 = moves all four extremities)
y/o	years old
1-	first degree, primary
2-	second degree, secondary
3-	third degree, tertiary

Index

About the Author

Jeffrey A. Pike's educational background includes studies at the Polytechnic Institute of New York, New York University, and the University of Michigan. His professional credentials include the following: Senior Technical Specialist, Ford Motor Company (retired after 32 years); Adjunct Professor, Wayne State University, Biomedical Engineering Department; Fellow, SAE International; Principal Lecturer and Organizer, SAE seminars on automotive safety (22 years) and forensic biomechanics (4 years); and invited speaker at conferences in the United States (including two White House conferences), Europe, China, India, Korea, and Australia.

Mr. Pike has served as a consultant to various agencies, including the Centers for Disease Control (CDC), National Academy of Sciences, National Highway Traffic Safety Administration, and state and local governments.

Mr. Pike's publications include technical papers, book chapters, and two previous books (*Automotive Safety* and *Neck Injury: The Use of X-Rays, CTs, and MRIs to Study Crash-Related Injury Mechanisms*), both of which are used as graduate course texts. His additional teaching experience includes guest lecturer at the Massachusetts Institute of Technology, the Medical College of Wisconsin, the University of Virginia, and Harvard Medical School.

In 2006, Mr. Pike retired from Ford Motor Company and founded Biomechanics Consulting, Inc. (BCI), which specializes in forensic and research studies of injury mechanisms and countermeasures. In addition to his consulting practice, Mr. Pike continues to actively participate on various committees and to lecture in numerous venues, including professional seminars, university classes, and developing countries. As he has for the past several decades, Mr. Pike continues to help train and mentor the next generation of safety professionals.